Water Hyacinth: A Potential Lignocellulosic Biomass for Bioethanol

Anuja Sharma · Neeraj K. Aggarwal

Water Hyacinth: A Potential Lignocellulosic Biomass for Bioethanol

Springer

Anuja Sharma
Department of Microbiology
Kurukshetra University
Kurukshetra, Haryana, India

Neeraj K. Aggarwal
Department of Microbiology
Kurukshetra University
Kurukshetra, Haryana, India

ISBN 978-3-030-35631-6 ISBN 978-3-030-35632-3 (eBook)
https://doi.org/10.1007/978-3-030-35632-3

This Springer imprint is published by the registered company Springer Nature Switzerland AG
The registered company address is: Gewerbestrasse 11, 6330 Cham, Switzerland

Contents

Abbreviations

%	Percent
<	Smaller than
>	Greater than
°C	Degree celsius
ABTS	2,2′-azino-bis(3-ethylbenzo-thiazoline-6 sulphonic acid)
AFEX	Ammonia fibre explosion
CBP	Consolidated bioprocessing
CI	Crystalline index
CMCase	Carboxymethyl cellulase
DMC	Direct microbial conversion
DP	Degree of polymerization
FPase	Filter paperase
FPU	Filter paper unit
FTIR	Fourier transformed infrared spectroscopy
g/g	Gram per gram
g/l	Gram per litre
gds	Gram dry substrate
GHS	Glycoside hydrolase
HBT	1-Hydroxybenzotriazole
HMF	Hydroxy methyl furfural
IFF	Integrated fungal fermentation
IMC	Initial moisture content
IU	International unit
kDa	kilodalton
Lac	Laccase
LHW	Liquid hot water
LiP	Lignin peroxidase
LMS	Laccase-mediated system
mg	Milligram
mg/ml	Milligram per millilitre

min	Minute
mm	Millimetre
MnP	Manganese peroxidase
MPa	Megapascals
MWL	Milled wood lignin
NMR	Nuclear magnetic resonance
pH	Potential of hydrogen
PKL	Polymeric kraft lignin
POL	Polymeric organosolv lignin
SEM	Scanning electron microscopy
SHF	Separate hydrolysis and fermentation
SiSF	Simultaneous saccharification and fermentation
SmF	Submerged fermentation
sp	Species
SSF	Solid state fermentation
U/g	Unit per gram
U/gds	Unit per gram dry substrate
U/l	Unit per litre
v/v	Volume/Volume
w/v	Weight/Volume
WHB	Water hyacinth biomass
XRD	X-ray diffraction

Chapter 1
Introduction

Abstract Scarcity of non-renewable energy sources due to the continuous depletion of fossil fuels has paved need for sustainable and environmentally friendly biofuels from biomass. Of all the natural resources used for production of biofuel, lignocellulosic biomass is the most attractive source due to its sustainability, wide abundance and easy availability. Efficient and cost-effective conversion of lignocellulosic biomass to biofuel production requires various steps including pretreatment, saccharification and fermentation process. In the present chapter, we have consolidated different generations of biofuels, composition of lignocellulosic biomass and a brief overview of the current status of biofuel production.

Keywords Bioethanol · Biofuel · Cellulose · Hemicellulose · Lignin · Lignocelluloses

The most important factors responsible for the increased demand of research and development in non-petroleum-based bioenergy sector are constant rise in energy demand, fast depleting resources and increased CO_2 emissions and air pollution. Biofuels as a replacement for fossil-based transport fuels are a viable and sustainable option since unlike fossil fuels which on combustion releases CO_2 that was captured million years ago, CO_2 released during the utilization of a biofuel is balanced by CO_2 that was captured during the growth of the biomass used for biofuel production, leading to a far less net impact on greenhouse gas levels. Thus, besides the obvious environmental benefits, the use of renewable raw materials to replace fossil fuels will have tremendous economic benefits, including the reduction of the crude oil dependency, trade deficit reduction, development of a strong biomass industry and the strengthening of agricultural markets. Biomass is an abundant resource including a wide variety of industrial, agricultural and forest resources available, which could provide suitable raw material for biofuel production. To be able to cover our global energy needs, the chosen renewable resources must have high energy content, be available in large amount, should be easily accessible and at a low cost. Without the fulfilment of these four requirements, the economic viability of the proposed renewable resource is questionable. Biofuels are generally classified as first-, second- and third-generation biofuels based on their source and production technology.

A. Sharma and N. K. Aggarwal, *Water Hyacinth: A Potential Lignocellulosic Biomass for Bioethanol*, https://doi.org/10.1007/978-3-030-35632-3_1

1.1 First-Generation Biofuel

First-generation bioethanol is produced from starch or sugar found in wide variety of agricultural crops by fermentation. The bioethanol produced by fermentation of sugars from agricultural crops such as sugarcane, sugar beet, sorghum and whey; and starchy feedstocks such as grains, viz. maize, wheat, and root crops such as cassava is some common examples of first-generation bioethanol [1]. First-generation bioethanol has played a significant role in laying a foundation for the establishment of the infrastructure and policy drivers required to support biofuels as renewable transport fuels in the international market. However, major limitations associated with first-generation biofuels include competition for the use of arable land with food and fibre crops, restricted market structures, absence of well-established agricultural practices in developing economies, high cost due to water and fertilizers requirements and threat to biodiversity.

1.2 Second-Generation Biofuel

Second-generation biofuels are produced from non-food crops such as biowastes and lignocellulosic biomass. Lignocellulosic materials are one of the world's largest renewable biochemical resources with an estimated annual production of 10–50 billion tons which may be available for large-scale biofuel production [2]. When compared to feedstock used for first-generation biofuel production, lignocellulosic biomass as a source for biofuel is a non-edible source that does not compete with food crops, can be cultivated specifically for bioenergy production leading to high productivity and increases land-use efficiency and soil quality being an above-ground plant material. Lignocellulosic biomass is, therefore, considered as the only foreseeable, feasible and sustainable resource for renewable fuel. A typical lignocellulosic biomass is mostly composed of about 40–50% cellulose, a glucose polymer; 25–35% hemicellulose, a heteropolymer made up of pentoses; 15–20% lignin, a non-polysaccharide made up of phenyl-propane units [3].

The production of biofuel from lignocellulosic biomass occurs in three major steps involving pretreatment of biomass by physical, chemical or biological methods, enzymatic hydrolysis of polysaccharides cellulose and hemicellulose into monomeric sugars and finally fermentation of these simple sugars to alcohol. Since lignin is not a polysaccharide and cannot be hydrolyzed and fermented to produce bioethanol, it can be recovered after pretreatment step and used to produce heat and electricity at the bioethanol production facility for complete utilization of lignocellulosic biomass [4]. Conventional sources of lignocellulosic biomass include agricultural residues, dedicated herbaceous, hardwoods and softwoods while non-conventional sources include industrial cellulosic waste, municipal solid waste and weeds. Examples of second-generation feedstocks are corn stover, sugarcane bagasse, wheat straw, rice straw, food waste, cashew apple bagasse, etc.

Third-generation biofuels exploit algae as feedstock for biofuel production. Extensive research is going on in this field to engineer algae in an attempt to increase biofuel production as well as to study algal biodiversity in order to reveal highly competent fuel-producing species.

1.3 Chemical Composition of Lignocelluloses

1.3.1 Holocellulose

Cellulose and hemicelluloses together are known as holocellulose and usually account for 65–70% of the lignocellulosic biomass. These polymers are made up of simple sugars (hexoses and pentoses) like D-glucose, D-galactose, D-mannose, D-xylose, L-arabinose, D-glucuronic acid, L-rhamnose and D-fucose.

1.3.1.1 Cellulose

Cellulose, the major chemical component of the fibre wall and a homopolysaccharide, is composed entirely of glucose (D-glucopyranose) units linked together by β-1,4-glycosidic bonds (β-D-glucan) [5]. Cellulose, the largest natural polymer in nature, is widespread in both primitive and highly evolved plants. Although the size of a cellulose molecule is generally defined and calculated as its degree of polymerization (DP) which is the number of glucose units present in a single chain, the conformational analysis indicated that cellobiose and not glucose is the basic structural unit of cellulose molecule [6]. Cellulose has a linear structure and a strong tendency to form intra- or intermolecular hydrogen bonds [7]. These hydrogen bonds along with Vander Waal's forces make cellulose structurally so complex with the individual cellulose molecules arrayed in bundles called microfibrils which promote aggregation into crystalline, highly ordered regions. Each microfibril contains approximately 40 individual cellulose molecules [8]. Within these microfibril bundles, the cellulose is highly ordered and thus appears crystalline when subjected to X-ray diffraction for measurements. Arrangement of fibrils and microfibrils of cellulose molecule is shown in Fig. 1.1. In addition to the crystalline region, there are a small percentage of non-organized cellulose chains, which form amorphous region of cellulose. This arrangement of crystalline and amorphous regions of cellulose molecule is responsible for imparting its interesting properties of stiffness and rigidity on the one hand and flexibility on the other hand. Crystallinity index (CI) is the parameter used to determine the relative amount of crystalline material in cellulose. The CI of celluloses can be measured using various techniques including XRD, solid-state ^{13}C NMR, infrared (IR) spectroscopy and Raman spectroscopy. The CI of cellulose has also been used to interpret structural changes in the cellulose molecule after physicochemical and biological pretreatments of lignocellulosic biomass. However, it has

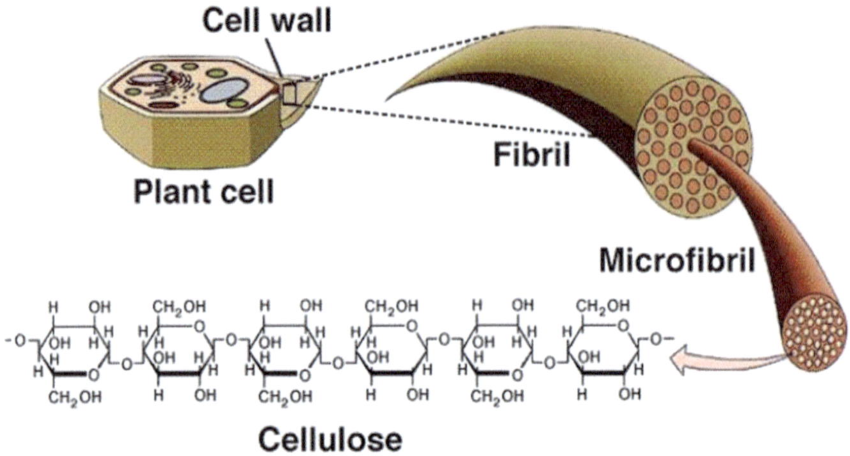

Fig. 1.1 Arrangement of fibrils, microfibrils and cellulose in plant cell wall [11]

been found that CI of cellulose varies (39–67% of Avicel cellulose) depending on the source of cellulose and technique used for measurement [9]. It is generally stated that amorphous regions in a partially crystalline cellulose are hydrolyzed first followed by crystalline domains, thus resulting in an increased crystallinity index and decreased rate of hydrolysis at later stages of hydrolysis. However, the correlation between crystallinity of cellulose and rate of hydrolysis is difficult to establish due to the use of different types of cellulose used in different studies [10]. The enzymes involved in cellulose degradation belong predominantly to hydrolases like cellulase (endoglucanase), 1, 4-β-cellobiosidase and β-glucosidase which cleave the glycosidic bonds [5].

1.3.1.2 Hemicelluloses

Hemicelluloses are complex branched heteropolymer composed of different pentoses such as D-xylose, D-arabinose and hexoses such as D-mannose, D-glucose and D-glucuronic acid. Hemicellulose has branches with short lateral chains consisting of different sugars, sugar acids and acetyl esters. These acid and ester groups render hemicelluloses their non-crystalline or poorly crystalline nature so that they exist more as a gel than as fibres [12]. Hemicelluloses also have a lower degree of polymerization and are more easily hydrolyzed than cellulose [13]. Hemicellulose is more soluble than cellulose and is frequently branched with degree of polymerization of 100–200. Hemicellulose is classified on the basis of the monomeric sugar present in the backbone of the polymer, e.g. mannan (β-1,4-linked mannose) or xylan (β-1,4-linked xylose) hemicelluloses with xylan hemicellulose being the most abundant in nature (Fig. 1.2). Hemicellulose generally falls into four major classes: (a) unbranched chains such as (1-4)-linked xylans or mannans; (b) helical chain such as

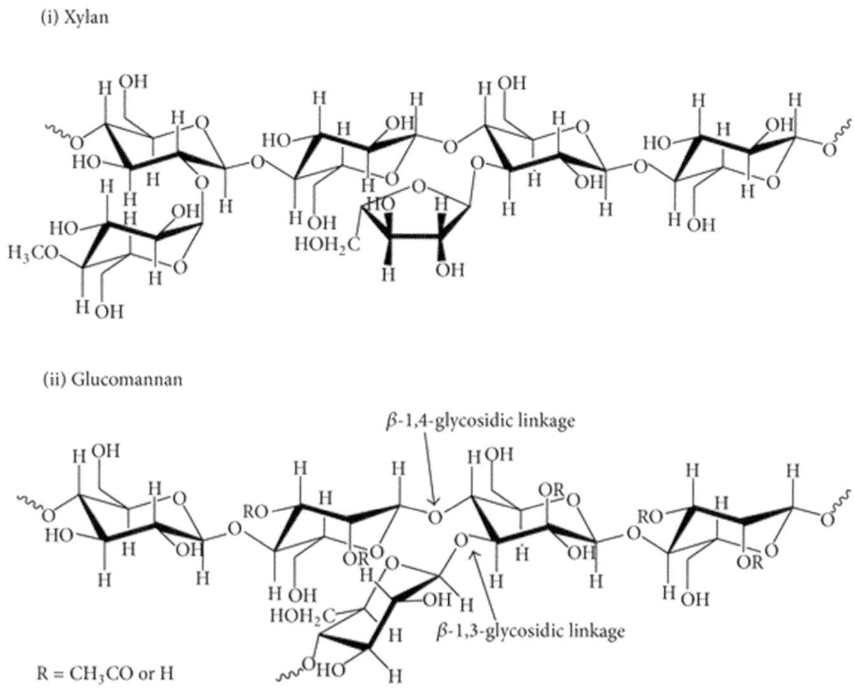

Fig. 1.2 Structure of hemicellulose [15]

(1-3)-linked xylans; (c) branched chains such as (1-4)-linked galactoglucomannans; and (d) pectic substances such as polyrhamnogalacturonans [6]. In the branched chain type hemicellulose like galactoglucomannan, the main chain of glucose and mannose residues is connected with β-(1, 4) glycosidic bond with the side chain attached to the main chain via α-(1, 6) bonds. The hemicellulose from hardwood and agricultural residues is typically rich in xylan accounting for about 30–35% of total dry weight while softwood mostly contains mannan and less xylan [14]. Xylan can be hydrolyzed to its simple form xylose by enzymes endo-1,4-β-xylanase (EC 3.2.1.8) and 1,4-β-xylosidase (EC 3.2.1.37). The degradation of hemicellulose is known to be more common in fungi than bacteria [5].

1.3.2 Lignin

Lignin is one of the most complex and recalcitrant natural polymers with regard to its chemical structure and composition. It is a cementing, tough noncarbohydrate polyphenolic structural constituent of cell wall of all the vascular plants, interspersed with the hemicelluloses and thus keeping cellulose locked within structural confinements of cell wall. Lignin is extremely resistant to enzymatic degradation due to its structural complexity provided by the oxidative coupling of monolignols (lignin monomers, monomeric precursors), the three primary hydroxycinnamyl alcohols: *p*-coumaryl, coniferyl and sinapyl alcohols as shown in Fig. 1.3 [16]. The corresponding phenylpropanoid units in the lignin polymer are known as *p*-hydrophenyl (H), guaiacyl (G) and syringyl (S) units, respectively, based on the methoxy substitution on the aromatic rings [17]. Its resistance to microbial degradation is also attributed to the extensive cross-linking between its polyphenolic constituents and biologically stable ether linkages. Lignin is responsible for providing the compressional strength to the plant cell wall while cellulose provides the flexible strength to the plant. Ester linkages between the free carboxyl group of hemicellulose and the benzyl groups of lignin molecule lead to the formation of the lignin–carbohydrate complex (LCC) which embeds the cellulose thus providing it with the resistance against microbial and chemical degradation [5].

Highest concentrations of lignin are usually present in the middle lamellae (intercellular) region of the plant where the lignin holds the plant cells together, thereby providing strength and rigidity to the plant, conferring structural support, impermeability and resistance against microbial attack and oxidative stress [18]. During biosynthesis of lignin, the precursors are transformed through enzymatic dehydrogenation reactions to phenoxy radicals which then are polymerized to form the final

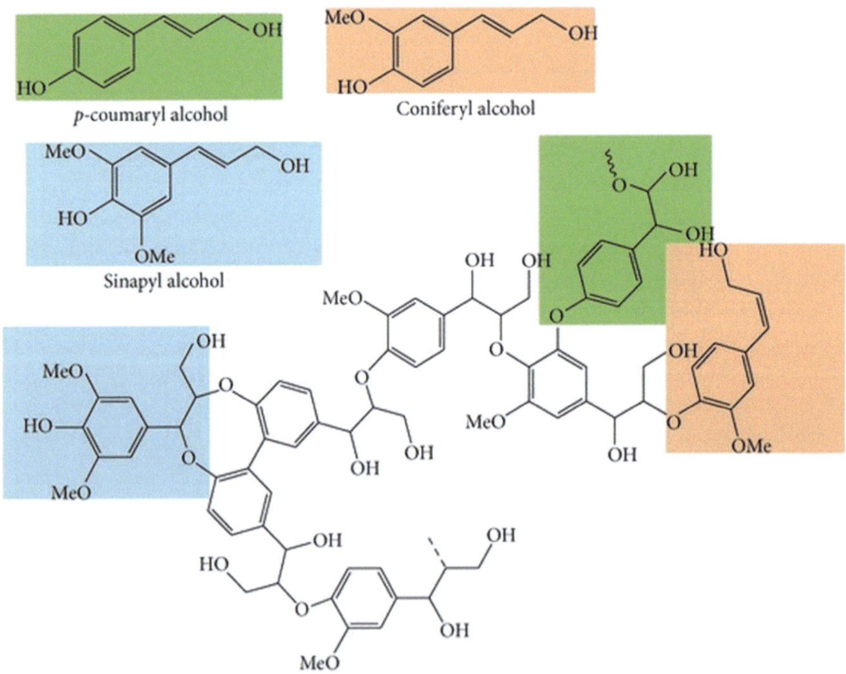

Fig. 1.3 Chemical structure of lignin [19]

lignin structure. Plant peroxidases catalyze the one-electron oxidation of these precursors to generate phenoxy radicals which diffuse away from the enzyme to couple with one another. Bonds connecting lignin precursors together comprise 60–80% of ether linkages, of which the most common type is β-O-4-bonding. Rest part of linkages has been identified as carbon-carbon (C–C) bonds and ester (C–O–C) bonds [7]. Lignin has no optical activity in contrast to other compounds because the phenoxy radicals formed during enzymatic dehydrogenation process randomly couple with one another to form the lignin polymer.

1.4 Bioethanol and Status Worldwide

Bioethanol is produced through distillation of the ethanol wash obtained after the fermentation of sugars derived from various biomasses and can be utilized as a liquid fuel, either in neat form or in petrol blends. Ethanol has a high octane number which makes its blend achieve the octane boosting or anti-knock effect similar to petroleum-derived aromatics like benzene. One example is the blend E85 which consists of 15% gasoline and 85% ethanol and has a prevalent usage as alternative fuel because of its advantage over pure ethanol.

With rapid depletion of the world reserves of fossil fuels, bioethanol has emerged as one of the alternative liquid fuels in recent years with immense amount of research dedicated to the production of ethanol and its socioeconomic and environmental impact. The use of starchy materials for ethanol production trace back to the twelfth century in typical beer-producing countries like Ireland. However, it was only in the early years of the twentieth century that the use of alcohol as a fuel for various combustion engines, especially for automobiles, became known and later gained popularity. It was in the 1970s that various ethanol programmes were administered and the National Alcohol Fuels Commission was established to study the potential of ethanol-based fuels [20]. Further support was provided by Chrysler, Ford and General Motors in 1980 when they released a statement stating that ethanol with blends of up to 10% would be covered in their vehicles warranty. Interest in the use of biofuels worldwide has grown dynamically since then due to the rapid depletion of oil reserves, exponential rise in energy demand and concerns over climate change from greenhouse gas emissions.

Bioethanol market has grown considerably from less than a billion litres in 1975 to more than 65 billion litres in 2008 and could grow to exceed 125 billion litres by 2020 [21]. Also, the global biofuels supply since the year 2000 increased by a factor of 8% to contribute to 4% of the world's total transport fuels in 2015 [22]. This equalled to approximately 35 billion gallons of biofuels, consisting roughly of 3:1 of ethanol to biodiesel [22]. Majority of this supply was contributed by first-generation biofuels produced from sugar, starch, vegetable oil or animal fat. This significant rise is attributed to national biofuel policies promoting biofuels and blending mandates which foster greater utilization of biofuels and may partly insulate biofuels during times of oil price flux [23]. According to International Energy Agency [24], the total worldwide demand for oil is estimated to rise by 1% per year especially due to the increasing demand of energy in developing countries like India (3.9% per year) and China (3.5% per year). Major contributors are industries such as aviation, marine transport and heavy freight, where biofuels are conceived as the only practical and low-carbon alternative [24]. Also, the recent signing of an agreement by 191 countries to curb aviation pollution accentuates that there is notable market future for continued biofuel adoption. However, while the broad interest and potential market for biofuels exist, near-term plant construction and double-digit supply growth have decelerated reflecting structural challenges and policy uncertainty in major markets [22].

The USA was the world's largest bioethanol producer, accounting for about 47% of the global bioethanol production in 2005 and 2006 [25]. According to Energy Independence and Security Act of 2007, USA mandates the use of a minimum volume of biofuel in transportation industry, but it does not compel biofuel production [26]. The Environmental Protection Agency which oversees this act was essentially designed to increase the consumption of renewable fuel from 9 billion gallons in 2008 to 36 billion in 2022 [26]. Feedstock incentives are also in play to provide financial aid to set up biomass feedstock crops for advanced biofuels facilities.

The EU has also adopted a Biomass Action Plan that sets out sustainability requirements for biofuels encompassing a reduction in GHG emission, land management, measures to accelerate the development of bioenergy from agricultural crops, wood, agricultural and industrial wastes. EU Energy and Climate Change Package (CCP) 2009 outline the regulations for the use of transport-based biofuels [27]. The CCP includes requirements that envisage a 20% renewable energy mix in total energy consumption by 2020.

Brazil is the world's largest exporter of bioethanol and also the second largest producer after the USA. Brazil is one of the most developed nations in ethanol production with the blending requirement for ethanol recently between 18 and 27.5%, currently 27% [28]. A stepped timetable designated to increase the biodiesel mix from 7% to 1 0% by 2019 has also been designed. Tax exemptions and incentives are available for both ethanol and biodiesel, in order to encourage production and social inclusion. A reduced tax burden for ethanol mixed fuel vehicles versus gasoline-only fuelled vehicles is in place. According to the National Biodiesel Production Programme (PNPB) launched in 2004, suppliers are compelled to procure vegetable oil from small-scale producers and family farms [28].

The Planning Commission of the Government of India identified bioethanol and biodiesel as the principal biofuels to be developed for the nation in an extensive report on the development of biofuels in the year 2003. The Ethanol Blended Petrol Programme (EBPP) launched in January 2003 regulated 5% ethanol blending with petrol in the states of Andhra Pradesh, Goa, Gujarat, Haryana, Karnataka, Maharashtra, Punjab, Tamil Nadu, Uttar Pradesh and Uttaranchal and in the union territories of Daman and Diu, Dadra and Nagar Haveli and Chandigarh. Also, the Ministry of New and Renewable Energy (MoNRE) released a national biofuel policy in 2009 with a target to replace 20% of petroleum-based fuel with biofuels by the end of the 12th Five Year Plan in 2017 [29]. The government of India also announced a mandatory blending requirement of 10% ethanol in gasoline from the beginning of the October 2015/2016 sugarcane season.

The major goal for the country now is the amendment of the 2009 biofuel policy with the focus on development and utilization of indigenous lignocellulosic feedstocks with a thrust on research and development for efficient production of second-generation biofuels and a blending mandate for both bioethanol and biodiesel.

References

1. Burk MJ (2010) Sustainable production of industrial chemicals from sugars. Int Sugar J 112: 30–35
2. Wan C, Li Y (2012) Fungal pretreatment of lignocellulosic biomass. Biotechnol Adv 30(6):1447–1457
3. Holtzapple MT, Jun JH, Ashok G, Patibandla SL, Dale BE (1991) The ammonia freeze explosion process: a practical lignocellulose pretreatment. App Biochem Biotechnol 28:59–74
4. Larson ED (2008) Biofuel production technologies: status, prospects and implications for trade and development. In: United Nations conference on trade and development (UNCTAD)
5. Schmidt O (2006) Chapter 3: Physiology. Chapter 4: Wood cell wall degradation. In: Wood and tree fungi: biology, damage, protection, and use. Springer, New York
6. Ramos LP (2003) The chemistry involved in the steam treatment of lignocellulosic materials. Quim Nova 26(6):863–871
7. Harinen S (2004) Analysis of the top phase fraction of wood pyrolysis liquids. In: Master's Thesis, Department of Chemistry, Laboratory of Applied Chemistry, University of Jyvaskyla
8. Zhao X, Song Y, Liu D (2011) Enzymatic hydrolysis and simultaneous saccharification and fermentation of alkali/peracetic acid-pretreated sugarcane bagasse for ethanol and 2,3-butanediol production. Enzyme Microb Technol 49:413–419
9. He J, Cui S, Wang SY (2010) Preparation and crystalline analysis of high grade bamboo dissolving pulp for cellulose acetate. J Polym Sci 107:1029–1038
10. Hall M, Bansal P, Lee JH, Realff MJ, Bommarius A (2010) Cellulose crystallinity—a key predictor of the enzymatic hydrolysis rate. FEBS J. https://doi.org/10.1111/j.1742-4658.2010.07585.x
11. Cankaya N (2015) Cellulose grafting by Atom transfer Radical polymerization method. In: Cellulose Fundamental aspects and current trends. INTECH. http://dx.doi.org/10.5772/61707
12. Kirk TK, Cullen D (1988) Enzymology and molecular genetics of wood degradation by white-rot fungi. Environmentally friendly technologies for the pulp and paper industry. John Wiley and Sons, NewYork, pp 273–307
13. Zaldivar J, Nielsen J, Olsson L (2001) Fuel ethanol production from lignocellulose: a challenge for metabolic engineering and process integration. App Microbiol Biotechnol 56:17–34
14. Kuhad RC, Gupta R, Khasa YP (2011) Bioethanol production from lignocellulosics: an overview. In: Wealth from waste, 3rd edn. TERI Press, New Delhi, India
15. Scheller HV, Ulvskov P (2010) Hemicelluloses. Annu Rev Plant Biol 61:263–289
16. Vivekanand V, Dwivedi P, Sharma A, Sabharwal N, Singh RP (2008) Enhanced delignification of mixed wood pulp by *Aspergillus fumigatus* laccase mediator system. World J Microbiol Biotechnol 24:2799–2804
17. Higuchi T (2006) Look back over the studies of lignin biochemistry. J Wood Sci 52:2–8
18. Oluwdare AO, Asagbara EO (2008) Biodegradation of *Sterculia setigera* (Sterculiaceae) chips and its effects on wood basic chemical composition. Inter J Bot 4(4):461–465
19. Lee HV, Hamid SB, Zain SK (2014) Conversion of lignocellulosic biomass to nanocellulose: structure and chemical process. Sci World J. https://doi.org/10.1155/2014/631013

20. Lansing (1983) Alcohol Fuels in Michigan. Department of Agriculture, State of Michigan, pp 3–4
21. Demirbas A (2007) Producing and using bioethanol as an automotive fuel. Energy Sour Part B 2:391–401
22. Renewable Energy Network 21 (REN21) (2016) Global status report. REN21, Paris, France
23. Annual Energy Review. Energy Information Administration (EIA). http://www.eia.gov/totalenergy/data/annual/index.php. Accessed 22 Nov 2018
24. Innovation Outlook: Advanced Liquid Biofuels (2016) International Renewable Energy Agency (IRENA). Abu Dhabi, UAE
25. Balat M, Balat H (2009) Recent trends in global production and utilization of bio-ethanol fuel. Appl Energy 86:2273–2282
26. Bramcourt K (2016) The renewable fuel standard (RFS). In: Brief: congressional research service. Washington, DC, USA, pp 7–5700
27. Araujo K, Mahajan D, Kerr R, Silva M (2017) Global biofuels at the crossroads: an overview of technical, policy, and investment complexities in the sustainability of biofuel development agriculture. https://doi.org/10.3390/agriculture7040032
28. US Department of Agriculture (USDA), Brazil Biofuels Annual (2016) GAIN report number BR 16009. Brazilian law 13.263/2016. USDA, Washington DC, USA
29. US Department of Agriculture (USDA), India Biofuels Annual (2016) GAIN report number IN 6088. USDA, Washington DC, USA

Chapter 2
Water Hyacinth: An Environmental Concern or a Sustainable Lignocellulosic Substrate

Abstract *Eichhornia crassipes* is one of the world's most pernicious aquatic weeds owing to its expeditious proliferation rate, ecological adaptability and survival strategies and deleterious impact on environment, human health and socio-economic development. A number of weed control methods including physical/mechanical removal, chemical methods and biological control agents have been used to eradicate or manage it worldwide. However, due to various environmental and financial challenges associated, none of these strategies or their combinations has been effective in completely eradicating this noxious weed. On the contrary, water hyacinth has demonstrated its potential in various biotechnological applications like bioremediation and bioadsorption of metal from polluted aquatic environment; bioenergy production, composting and vermicomposting, as animal and fish feed; as carbon source for microbial growth; various medicinal and other uses. In this chapter, different applications of water hyacinth along with its impact on environment and various control methods have been discussed in brief.

Keywords Biofuel · Biocontrol · *Eichhornia* · Lignocellulose · Water hyacinth · Weed

Eichhornia crassipes commonly known as water hyacinth is a native of Amazon River in South America and one of the world's most noxious aquatic weeds. Due to its highly invasive and detrimental nature, it has been inducted in 100 of the world's worst invasive alien species, a selection from the global database [1, 2]. This aquatic macrophyte is a monocotyledon belonging to the family Pontederiaceae (pickerelweed) and order commelinales. The presence of spectacular violet and yellow flowers in bunches and bulbous green leaves in this free-floating aquatic plant made it an attractive ornamental plant. For similar reasons, this tropical species was introduced to Australia for the first time as an ornate plant to decorate botanical gardens and ornamental ponds have since invaded more than 50 countries and are pervasive in Central America, Central and Western Africa, South-eastern USA and Southeast Asia [3]. Its rapid proliferation rate [1] and extraordinary adaptability to extreme conditions and efficient survival strategies contribute to its high degree of invasion. Once present in an ecosystem, it can spread at extremely high rates of up to 100–120 t of dry biomass per hectare per year and can easily

A. Sharma and N. K. Aggarwal, *Water Hyacinth: A Potential Lignocellulosic Biomass for Bioethanol*, https://doi.org/10.1007/978-3-030-35632-3_2

wipe out the entire domestic vegetation [1, 4]. Its dominance in the tropics and subtropics is attributed to irregular wastewater management and poor sanitation making it even more difficult to control [4]. Water hyacinth expands in the form of dense coverage facilitated by its complex root system [5]. This impenetrable coverage blocks waterways and interferes with drainage system, navigation and fishing industry, limits the penetration of light, depletes dissolved oxygen and other nutrients, thus causing destruction to the aquatic ecosystem [5]. A number of strategies for management of water hyacinth are being developed and implemented throughout the world but none have been effective enough to eradicate this vicious weed completely. Thus, its control or management is still dependent on methods that limit the ecological devastation and socio-economic damage caused by it.

2.1 Origin

Water hyacinth is originally from the rain forests of Amazon River, Brazil, but grows naturally in other parts of South America. Its infestation in the river Nile was recorded in the late eighteenth century [6]. Water hyacinth's initial introduction as an ornamental aquatic plant for botanical gardens and ponds dates back to the nineteenth century [7]. Its proliferation throughout the freshwater rivers and lakes of tropical and subtropical regions taking over the entire native aquatic community is reported from the early twentieth century. Currently, it is the most troublesome weed of Africa, Asia, Australia, Egypt, India, Java, Central America, North America and New Zealand.

2.2 Ecology

Water hyacinth is one of the fastest-growing free-floating aquatic plants capable of growth in extreme environmental conditions which justify its high degree of invasion around the world. It can flourish well in both tropical and temperate conditions with a wide range of growth temperature and an optimum of 28–30 °C [8]. However, below freezing temperatures may lead to the freezing of the tip of rhizome inhibiting the growth of water hyacinth and ultimately the death of the plant. Its pH preference for growth varies from a mild acidic to mild alkaline range of 4.0–8.0. Water hyacinth is euryhaline and can grow in fresh and marine water but stagnant freshwater is most suitable for propagation [7, 8]. The plant is known for its high tolerance to extreme fluctuations in climatic and water conditions, temperature, pH and nutrient limitations including high salinity levels (0.24%), survival on damp soil and mild frost and presence of pollutants and toxic compounds in the water body [8]. Although water hyacinth can survive in nutrient limiting environment, its growth and biomass accumulation, ramet production, shoot: root ratio and plant height are greatly enhanced in water containing high concentration of nitrogen, phosphorus and potassium [8–11].

Another significant feature contributing to its high degree of invasion is that it can store nutrients for later stages of the life cycle [12].

2.3 Biology

E. crassipes is free-floating plant that can rise above the water surface from a few inches to a metre. Leaves are broad (10–20 cm), thick, bulbous and glossy in appearance with circular blades and are borne on a long and spongy stalk [13]. A rosette of six to ten leaves floating above the water surface and attached to a rhizome are present in a single plant. A stalk produces a single stalk that bears a spike of 8–15 attractive purple to pink flowers with six petals. The plant has a well-developed fibrous root system composed of unbranched roots with a conspicuous root cap [1]. Colour of roots in water hyacinth depends on their position and is purple black for free-hanging roots and white for roots fixed in soil [14]. Presence of foliar plasticity is another important feature of water hyacinth.

Water hyacinth can reproduce by both sexual and asexual means. Sexual reproduction takes place by formation of seeds while vegetative or asexual reproduction occurs by budding and stolen production. Both sexual and vegetative reproductions are efficient in producing a large number of individuals in a short period but the main mode of reproduction is vegetative [15]. During vegetative reproduction, daughter plant or ramets are produced on the stolen. Under favourable conditions of warm temperature, eutrophic waters and nutrient availability, the vegetative propagation is rapid with a doubling time of 11 to 18 days and the mat coverage can grow by as much as 60 cm/month. This extensive reproductive potential of water hyacinth along with fast proliferation causes reinfestation and rapid coverage on previously treated areas, making the process of control inefficient. Germination of seeds into mature plants occurs in a few days, depending on environmental factors and is quite sensitive to unfavourable conditions like oxygen stress, light and low temperature, and dormancy periods such as droughts [15–17]. Seed capsules each containing approximately 50 seeds can sink and remain dormant until periods of stress for as long as 20 years. Upon attaining the favourable conditions of growth and reproduction, these seeds can germinate and renew the growth cycle. During flowering, the plant produces inflorescences having up to 23 flowers 10–15 weeks after germination. [15].

2.4 Impacts of Water Hyacinth

Water hyacinth exhibits deleterious effects on aquatic environment and ecosystem, water quality, human health and socio-economic development of the region infested by it. Major problems caused by it are due to its uncontrollable rapid extension as condensed mats which congest the water bodies entirely. This impenetrable coverage blocks waterways of rivers and canals causing obstruction in navigation, fishing, recreation, irrigation and hydroelectric power generation and may even lead to flooding [18]. *Eichhornia* degrades the water quality due to higher siltation and sedimentation within the plant's complex root structure and decomposing plants. Higher evapotranspiration from water hyacinth leaves covering the water surface is another factor which influences water level [19]. Decrease in the dissolved oxygen concentrations is another implication creating favourable breeding conditions for mosquitoes and leading to increased occurrence of waterborne diseases like malaria, encephalitis and filariasis [20]. Water hyacinth stabilizes the level of both pH and temperature and thus influences stratification within the lotic system.

Water hyacinth is known to alter the aquatic habitat and has the capacity to overgrow and replace the domestic vegetation and associated fauna causing an imbalance in aquatic ecosystem [21]. The productivity of phytoplankton is greatly influenced by water hyacinth infestation with an initial rise in certain colonial types entangled within the complex roots followed by an overall reduction. A decrease in population and diversity of fish community accompanied by inaccessible fishing sites is another negative impact of water hyacinth invasion [5]. However, the range of impact depends on various factors like initial composition of fish, phytoplankton and vertebrate community, and oxygen and nutrient concentration [22]. Some examples of water bodies heavily infested with water hyacinth and facing problems like interference with irrigation, navigation, freshwater supply and fishing industry, increased cases of waterborne diseases, eradication of natural vegetation from different parts of the world includes Lake Victoria, East Africa; Yamuna River, Delhi, India; Lake Navishka; Lake Chapala, Mexico; and Kafue River, Zambia [23].

2.5 Control

Water hyacinth has become a major problematic weed due to its unstoppable and rapid growth and needs to be managed. A large number of individual and combined methods are being used to eradicate it and billions of dollars are being wasted every year in control costs and economic losses. These weed management methods include physical/mechanical removal, use of chemicals and biological control agents and combination of two or more of these methods. Each of these individual methods has certain advantages but is not efficient enough because of various limitations associated with it. Physical methods may involve manual removal by harvesting; in situ cutting; mechanized removal using cranes, draglines, mowers, dredges, barges; installation of floating barriers to forestall the movement to other areas. Physical control has the advantage of no water use restriction. However, it is labour-intensive, decreases dissolved oxygen concentration, causes eutrophication and is a costly affair due to expensive cutting and dredging equipment [24, 25]. Chemical control requires the use of chemicals like glyphosate, diquat, 2,4-D amine, 2,4-dichlorophenoxyacetic acid, (2,4-D) + complexed copper, endothall dipotassium salt and endothall dimethylalkylamine salts [26, 27]. It is cost-effective as compared to the physical methods but is less selective, has deleterious effects on non-target algae and macrophytes and also leads to deoxygenation. Methods of biological control involve the use of insects like *Neochetina eichhorniae, N. bruchi* and *Sameodes albiguttalis*; fungal pathogens like *Alternaria eichhorniae*, *Alternaria alternata, Drechslera hawaiiensis* and *Ulocladium atrum* and allelopathic plants [28, 29]. Though the process is not labour and equipment-intensive and also has the potential to be self-sustaining, it is a slow process and insufficient reduction and resurgence in growth have also been observed.

A study investigated an integrated control method involving the use of glyphosate at low concentration with *A. alternata* as biological control agent [30]. Also, integrated weed management system comprising of chemical herbicides, microbial herbicide and arthropods has also been suggested. Initial strategies emphasized complete eradication of this notorious weed but due to the various limitations associated, steps to manage its density to levels that minimize its impact on ecological and socio-economic aspects are now being taken.

2.6 Various Utilities of Water Hyacinth

The potential of water hyacinth biomass to be used as a raw material in various applications provides an attractive opportunity for its management by its large-scale utilization. Water hyacinth contains a low lignin and high cellulose and hemicellulose content making it is an attractive source of biomass. The cellulose and hemicellulose contents of water hyacinth range from 17.8 to 34.19% and 17.66 to 49.2%, respectively, and the lignin content ranges from 1.9 to 26.36% [31–33]. The average chemical composition of water hyacinth is detailed in Table 2.1. This variation in chemical composition of water hyacinth in different studies can be attributed to

Table 2.1 Chemical analysis of water hyacinth [49]

Constituent	Value (%)
Cellulose	21.50
Hemicellulose	33.90
Lignin	7.01
Ash	12.10
Fat	1.65
Crude protein	13.75
Total solids	5.01
Carbon	45–50
Hydrogen	5.3–5.5
Nitrogen	1.8–3.2
Sulphur	0.25–0.35

the variation in age of plant at harvesting, environmental conditions of growth and location from where the water hyacinth was collected. High biomass productivities and abundant availability of this plant offer another advantage resulting for large amount of utilizable biomass. Another benefit associated with water hyacinth is that it is an aquatic plant and thus does not compete with land resources [34]. Various studies have indicated water hyacinth as promising biomass for the production of bioethanol, biogas, animal and fish feed, compost and other valuable products as described in Table 2.2 [5, 34–40]. Several studies have also proved water hyacinth biomass to be suitable for bioethanol production [41]. Additionally, bioethanol yields from water hyacinth are comparable to the agro-residues, thus making it a potential raw material for biofuels production [34, 41]. Among the various pretreatment methods used to treat water hyacinth biomass, the dilute acid pretreatment at high temperature and pressure is the most effective for hemicellulose degradation while retaining most of the cellulose for bioethanol generation [31, 39, 41 and 42]. The combination of biological and mild chemical pretreatment of water hyacinth has also been explored [43]. Both separated hydrolysis and fermentation (SHF) [44] and simultaneous saccharification and fermentation (SiSF) [45] techniques have been used to produce bioethanol from water hyacinth biomass. Metal contaminated water hyacinth biomass was subjected to hydrolysis by dilute sulphuric acid (1% v/v) at 110 °C for an hour followed by fermentation using *Saccharomyces cerevisiae* for bioethanol production and resulted in ethanol yield of 55.20% [46]. Biological pretreatment of water hyacinth to obtain lignin-free biomass for bioethanol production seems promising since it is both cost-effective and environment-friendly, unlike chemical methods. In accordance, biological pretreatment of water hyacinth under solid-state cultivation by *Phanerohaete chrysosporium* was carried out and various culture conditions were optimized for enhanced delignification [47]. Additionally, biodiesel production from water hyacinth has also been reported [48].

Table 2.2 Various utilities of water hyacinth

Product	References
Bioethanol	[31, 34, 41, 45, 50]
Biodiesel	[48]
Biogas	[13, 51]
Animal feed	[5, 51, 52]
Fish feed	[51]
Compost	[53, 54]
Carbon source	[51, 55]
Medicinal uses	[56, 57]
Others (furniture, paper, rope, baskets)	[51, 57]

References

1. Hill M, Coetzee J, Julien M, Center T (2011) Encyclopedia of biological invasions. In: Simberlof D, Rejmanek M (eds) Water hyacinth. University of California Press, Berkeley, CA, pp 689–692
2. Lowe S, Browne M, Boudjelas S, Poorte M (2000) 100 of the world's worst invasive alien species: a selection from the global invasive species database. IUCN/SSC Invasive Species Specialist Group (ISSG), Auckland, New Zealand, pp 1–12
3. Lu J, Wu J, Fu Z, Zhu L (2007) Water hyacinth in China: a sustainability science-based management framework. Environ Manage 40:823–830
4. Villamagna AM, Murphy BR (2010) Ecological and socio-economic impacts of invasive water hyacinth (*Eichhornia crassipes*): a review. Freshwater Bio 55:282–298
5. Malik A (2007) Environmental challenge *vis* a *vis* opportunity: the case of water hyacinth. Environ Int 33:122–138
6. Hill G, Waage J, Phiri G (1997) The water hyacinth problem in tropical Africa. In: Proceedings of the international water hyacinth consortium. Washington, DC, USA. 18–19 Mar 1997
7. Ojeifo M, Ekokotu PA, Olele NF, Ekelemu JK (2000) A review of the utilisation of water hyacinth control measures for a noxious weed. In: Proceedings of the international conference on water hyacinth. New-Bussa, p 183. 27 Oct–1 Nov 2000
8. Burton J, Oosterhout E, Ensbey R, Julien M (2010) Water hyacinth (*Eichhornia crassipes*): weed of national significance. Department of Primary Industries, NSW, Australia
9. Xie Y, Wen M, Yu D, Li Y (2004) Growth and resource allocation of water hyacinth as affected by gradually increasing nutrient concentrations. Aquat Bot 79:257–266
10. Reddy KR, Agami M, Tucker JC (1989) Influence of nitrogen supply rates on growth and nutrient storage by water hyacinth (*Eichhornia crassipes*) plants. Aquat Bot 36:33–43
11. Singh HD, Nag B, Sharma AK, Baruah JN (1984) Nutrient control of water hyacinth growth and productivity. In: Proceedings of the international conference on water hyacinth. Hyderabad, India, pp 243–263. 7–11 Feb 1983
12. Heard TA, Winterton SL (2000) Interactions between nutrient status and weevil herbivory in the biological control of water hyacinth. J Appl Ecol 37:117–127
13. Adegunloye DV, Olosunde SY, Omokanju AB (2013) Evaluation of ratio variation of water hyacinth (*Eichhornia crassipes*) on the production of pig dung biogas. Int J Biol Sci 2:44–48
14. Penfound WT, Earle TT (1948) The biology of the water hyacinth. Ecol Monogr 18:447–472
15. Barrett SCH (1980) Sexual reproduction in *Eichhornia crassipes* (water hyacinth). II. Seed production in natural populations. J Appl Ecol 17:113–124
16. Obeid M, Tag el Seed M (1976) Factors affecting dormancy and germination of seeds of *Eichhornia crassipes* (Mart.) Solms from the Nile. Weed Res 16:71–80
17. Center TD, Dray FA Jr, Jubinsky GP, Grodowitz MJ (1999) Biological control of water hyacinth under conditions of maintenance management: can herbicides and insects be integrated? Environ Manage 23:241–256
18. Epstein P (1998) Weeds bring disease to the East African waterways. Lancet 351:577
19. Gopal B (1987) Water hyacinth. Elsevier Science Publishers, Amsterdam, 471, ISBN-13: 9780444427069
20. Perna C, Burrows D (2005) Improved dissolved oxygen status following removal of exotic weed mats in important fish habitat lagoons of the tropical Burdekin River floodplain, Australia. Mar Pollut Bull 51:138–148
21. Mangas-Ramirez E, Elias-Gutierrez M (2004) Effect of mechanical removal of water hyacinth (*Eichhornia crassipes*) on the water quality and biological communities in a Mexican reservoir. Aquat Ecosyst Health Manage 7:161–168
22. Kateregga E, Sterner T (2009) Lake Victoria fish stocks and the effects of water hyacinth. J Environ Dev 18:62–78
23. Plummer ML (2005) Impact of invasive water hyacinth (*Eichhornia crassipes*) on snail hosts of schistosomiasis in Lake Victoria, East Africa. EcoHealth 2:81–86

24. Greenfield BK, Siemering GS, Andrews JC, Rajan M, Andrews SP, Spencer DF (2007) Mechanical shredding of water hyacinth (*Eichhornia crassipes*): effects on water quality in the Sacramento-San Joaquin river Delta, California. Estuaries Coasts 30:627–640
25. Uka UN, Chukwuka KS, Daddy F (2007) Water hyacinth infestation and management in Nigeria inland waters: a review. J Plant Sci 2:480–488
26. Arora J, Mehra NK (2003) Species diversity of planktonic and epiphytic rotifers in the back-waters of the Delhi segment of the Yamuna River, with remarks on new records from India. Zool Stud Taipei 42:239–247
27. Sharma A, Aggarwal NK, Saini A, Yadav A (2016) Beyond biocontrol: water hyacinth-opportunities and challenges. J Environ Sci Technol 9(1):26–48
28. Wilson JRU, Ajuonu O, Center TD, Hill MP, Julien MH (2007) The decline of water hyacinth on Lake Victoria was due to biological control by *Neochetina* sp. Aquat Bot 87:90–93
29. Zhang MX, Ling B, Kong CH, Liang G, Dong YZ (2005) Allelopathic effects of lantana (*Lantana camera* L.) on water hyacinth (*Eichhornia crassipes* (Mart.) Solms). Allelopathy J 15:125–129
30. Ray P, Hill MP (2012) Impact of feeding by Neochetina weevils on pathogenicity of fungi associated with water hyacinth in South Africa. J Aquat Plant Manage 50:79–84
31. Nigam JN (2002) Bioconversion of water-hyacinth (*Eichhornia crassipes*) hemicellulose acid hydrolysate to motor fuel ethanol by xylose–fermenting yeast. J Biotechnol 97:107–116
32. Bergier I, Salis SM, Miranda CHB, Ortega E, Luengo CA (2012) Biofuel production from water hyacinth in the Pantanal wetland. Ecohydrol Hydrobiol 12(1):77–84
33. Ruan T, Zeng R, Yin XY, Zhang SX, Yang ZH (2016) Water hyacinth (*Eichhornia crassipes*) biomass as a biofuel feedstock by enzymatic hydrolysis. Bioresour 11(1):2372–2380
34. Bhattacharya A, Kumar P (2010) Water hyacinth as a potential biofuel crop. Elect J Environ Agric Food Chem 9:112–122
35. Klass DL, Ghosh S (1981) Methane production by anaerobic digestion of water hyacinth (*Eichhornia crassipes*). In: Fuel from Biomass and waters. Ann Arbor Science Publications Inc., MI, USA, pp 129–148
36. Isarankura-Na-Ayudhya C, Tantimongcolwat T, Kongpanpee T, Prabkate P, Prachayasittikul V (2007) Appropriate technology for the bioconversion of water hyacinth (*Eichhornia crassipes*) to liquid ethanol: future prospects for community strengthening and sustainable development. Exper Clin Sci Inter J 6:167–176
37. Carina CG, Cecilia MP (2007) Water hyacinths as a resource in agriculture and energy production: a literature review. Waste Manag 27:117–129
38. Hronich JE, Martin L, Plawsky J, Bungay HR (2008) Potential of Eichhornia crassipes for biomass refining. J Ind Microbiol Biotechnol 35:393–402
39. Kumar P, Barrett DM, Delwiche MJ, Stroeve P (2009) Methods for pretreatment of lignocellulosic biomass for efficient hydrolysis and biofuel production. Ind Eng Chem Res 48:3713–3729
40. Almoustapha O, Kenfack S, Millogo-Rasolodimby J (2009) Biogas production using water hyacinth to meet collective energy needs in a Sahelian country. Field Actions Sci Rep 2:27–32
41. Mishima D, Kuniki M, Sei K, Soda S, Ike M, Fujita M (2008) Ethanol production from candidate energy crops: Water hyacinth (*Eichhornia crassipes*) and water lettuce (*Pistia stratiotes* L.). Bioresour Technol 99:2495–2500
42. Sun Y, Cheng J (2002) Hydrolysis of lignocellulosic materials for ethanol production: a review. Bioresour Technol 83:111
43. Ma F, Yang N, Xu C, Yu H, Wu S, Zhang X (2010) Combination of biological pretreatment with mild acid pretreatment for enzymatic hydrolysis and ethanol production from water hyacinth. Bioresour Technol 101(24):9600–9604
44. Sornvoraweat B, Kongkiattikajorn J (2010) Separated hydrolysis and fermentation of water hyacinth leaves for ethanol production. KKU Res J 15:794–802
45. Kasthuri T, Gowdhaman D, Ponnusami V (2012) Production of ethanol from Water Hyacinth (*Eichhornia crassipes*) by *Zymomonas mobilis* CP4: optimization studies. Asian J Sci Res 5:285–289

46. Mahmood T, Malik SA, Hussain ST (2010) Biosorption and recovery of heavy metals from aqueous solutions by *Eichhornia crassipes* (water hyacinth) ash. BioResources 5:1244–1256
47. Sari E, Syamsiah S, Sulistyo H, Muslikhin (2011) The kinetic of biodegradation lignin in water hyacinth (*Eichhornia crassipes*) by *Phanerochaete chrysosporium* using Solid State Fermentation (SSF) method for bioethanol production, Indonesia. World Acad Sci Eng Technol 5:6–21
48. Sagar CV, Kumari NA (2013) Sustainable biofuel production from water hyacinth (*Eicchornia Crassipes*). Int J Eng Trends Technol 4:4454–4458
49. Deshpande SK, Bhotmange MG, Chakrabarti T, Shashtri PN (2008) Production of cellulose and xylanase by *Trichoderma resei* (QH 9414 mutant), *Aspergillus niger* and mixed culture by solid state fermentation of Water hyacinth (*Eicchornia crassipes*). Indian J Chem Technol 15:449–456
50. Das S, Gangly A, Deya A, Ting Y-P, Chatterjee PK (2014) Characterization of water hyacinth biomass and microbial degradation of the biomass under solid state fermentation using a Lignocellulolytic fungus (*Alterneria* sp. NITDS1). J Chem Bio Phys Sci 4(3):2279–2293
51. Jafari N (2010) Ecological and socio-economic utilization of water hyacinth (*Eichhornia crassipes* Mart Solms). J Applied Sci Environ Manage 14:43–49
52. Akinwande VO, Mako AA, Babayemii OJ (2013) Biomass yield, chemical composition and the feed potential of water hyacinth (*Eichhornia crassipes*, Mart. Solms-Laubach) in Nigeria. Int J AgriSci 3:659–666
53. Lata N, Veenapani D (2011) Response of water hyacinth manure on growth attributes and yield in *Brassica juncea*. J Central Eur Agric 12:336–343
54. Dhal GS, Singh WR WR, Khwairakpam M, Kalamdhad AS (2012) Composting of water hyacinth using saw dust/rice straw as a bulking agent. Int J Environ Sci 2:1223–1238
55. Gupta M, Sarmah R, Gupta S (2004) Nutrient composition of *Pleurotus sajor-caju* grown on different substrates. J Food Sci Technol 41:584–586
56. Lalitha P, Jayanthi P (2014) Antiaging activity of the skin cream containing ethyl acetate extract of *Eichhornia crassipes* (Mart.) solms. Int J PharmTech Res 6:29–34
57. Knoshaug EP, Shi B, Shannon TG, Mleziva MM, Pienkos PT (2013) The potential of photosynthetic aquatic species as sources of useful cellulose fibers-a review. J Appl Phycol 25:1123–1134

Chapter 3
Lignocellulolytic Enzymology

Abstract Lignocellulosic biomass as one of the most abundant, sustainable and cost-effective feedstocks for biofuel and other biochemical production has been quite challenging due to the natural recalcitrance of lignocellulose composed of lignin, cellulose and hemicellulose to enzymatic actions. The extreme recalcitrant nature of lignin has been the major hindrance during lignocellulose depolymerization leading to inefficient enzymatic conversion of the cellulose and hemicellulose fraction of lignocellulose to their sugar monomers for their further utilization in the production of biocommodities. The effective hydrolysis of lignocellulosic biomass requires the synergetic action of three major types of enzymes, viz. cellulases, hemicellulases and lignases (lignocellulolytic enzymes) with specific actions for complete deconstruction of the complex lignocellulosic structure. The present chapter discusses the origin, structure, source and mechanisms of these enzymes and other accessory enzymes involved in complete and efficient depolymerization of lignocellulose.

Keywords Cellulase · Hemicellulase · Laccase · Lignin · Peroxidase · Veratryl

The demand for development of cellulosic biofuel as an alternative fuel has specially spurted in the last decade due to various economic and environmental concerns associated with fossil fuels. Lignocelluloses as one of the most abundant and renewable biomasses for the production of cellulosic biocommodities especially biofuels due to its rich cellulose content have technical challenges because of the recalcitrance of lignocellulose to enzymatic degradation. A typical lignocellulosic substrate is mainly composed of cellulose (30–40%), hemicellulose (20–30%) and lignin (20–30%), but the actual composition varies with different feedstocks. The resistance towards lignocellulose depolymerization and enzymatic decomposition is accredited to various morphological and physicochemical variables such as recalcitrant nature of lignin, degree of crystallinity, degree of polymerization, hemicellulose sheathing and particle size of the substrate. The effective hydrolysis of lignocellulosic biomass is attained by application of different enzymes with specific roles in deconstruction of the complex lignocellulosic structure. The process involves the synergetic action of three major types of enzymes, viz. cellulases, hemicellulases and lignases (ligninolytic enzymes) and other accessory enzymes for complete degradation activity. A wide range of micro-organisms including bacteria, actinomycetes and fungi have

A. Sharma and N. K. Aggarwal, *Water Hyacinth: A Potential Lignocellulosic Biomass for Bioethanol*, https://doi.org/10.1007/978-3-030-35632-3_3

potential to produce these enzymes. Among these, fungi are widely known to be the most efficient producers of lignocellulolytic enzymes with some examples given in Table 3.1.

Table 3.1 Organisms with lignocellulolytic potential

Name of the organism	Cellulases	Hemicellulases	Ligninolytic enzymes	References
Ceriporiopsis subvermispora	–	Xylanase	MnP	[1]
Aspergillus fumigatus *Alternaria* sp. *Penicillium* sp. *Trichoderma* sp.	CMCase FPA	NA	Laccase, MnP	[2]
Trametes trogii	β-glucosidase, Endoglucanase	–	Laccase, MnP	[3]
Cerrena maxima *Pycnoporus coccineus* *Coriolopsis polyzona* *P. ostreatus* *T. versicolor* *F. trogii*	CMCase FPA	NA	Laccase, MnP	[4]
Cerrena unicolor *Trametes versicolor*	CMCase	–	Laccase, MnP	[5]
Pleurotus sajor-caju	CMCase	Xylanase	Laccase, MnP, LiP	[6]
A. niger *B. oryzae* *P. chrysosporium* *Magnaporthe grisea*	β-glucosidase, cellobiosidase	β-xylosidase α-arabinofuranosidase galactosidase	Laccase	[7]
Fusarium solani	CMCase FPA	Xylanase	Laccase, MnP, LiP	[8]
Trametes sp. 44	CMCase	Xylanase	Laccase, MnP, LiP	[9]
Alternaria alternate	CMCase	NA	Laccase	[10]
Pleurotus ostreatus *Lentinula edode*	CMCase	Xylanase	Laccase, MnP, LiP	[11]
Coprinus comatus *Trichoderma reesei*	CMCase	Xylanase	Laccase	[12]
Daedalea flavida	CMCase FPA	Xylanase	Laccase, LiP	[13]

3.1 Ligninolytic Enzymes

Three major groups of enzymes responsible for the degradation of lignin are lignin peroxidase (LiP; EC 1.11.1.14), manganese peroxidase (MnP; EC 1.11.1.13) and laccase (EC 1.10.3.2). Table 3.2 comprises the list of major cofactors and reactions of these enzymes. The process can further be intensified by the action of several accessory enzymes including aryl alcohol oxidase (veratryl alcohol oxidase; EC 1.1.3.7), cellobiose/quinone oxidoreductase (EC 1.1.5.1), cellobiose dehydrogenase (EC 1.1.99.18), glyoxal oxidase (EC 1.2.3.5) and pyranose 2-oxidase (glucose 1-oxidase; EC 1.1.3.4). In addition to their lignin-degrading potential, these enzymes have also been exploited in textile industry to decolourize textile effluents, bleach and synthesize dyes; in food industry to eliminate undesirable phenolics for soil bioremediation, and enhance the nutritive value of crops. A list of important organisms, both fungi and bacteria producing ligninolytic enzymes, is discussed in Table 3.3.

Table 3.2 Enzymes involved in lignin degradation and their main reaction

Enzyme	Cofactor or substrate mediator	Main reaction
Laccase, Lac	O_2, mediators (ABTS)	Phenols are oxidized to phenoxyl radicals, other reactions in the presence of mediators
Lignin peroxidase, LiP	H_2O_2, veratryl alcohol	Oxidation of aromatic ring cation radical
Manganese peroxidase, MnP	H_2O_2, Mn organic acids as chelator, thiols, unsaturated lipids	Oxidation of Mn(II) to Mn(III), chelated Mn(III) oxidizes phenolic compounds to phenoxy radicals

Table 3.3 Ligninolytic enzymes producing organisms

Name of the organism	Laccase	MnP	LiP	References
Fungi				
Phanerochaete chrysosporium	–	+	+	[14, 15]
Polystictus sanguineus *Phlebia radiata* *Phlebia tremellosa*	+	+	+	[16]
Rigidoporus lignosus	+	+	–	[17]
Coriolopsis rigida, Pleurotus laciniatocrenatus, Pycnoporus sanguineus, T. villosa, Trametes pavonia	+	+	–	[18]
A. fumigatus, Alternaria sp., *Penicillium* sp., *Trichoderma* sp., *Acremonium* sp.	+	+	–	[2]
P. pinophilum (TERI DB1), *A. flavus*	+	+	–	[19]

(continued)

Table 3.3 (continued)

Name of the organism	Laccase	MnP	LiP	References
F. verticillioides, A. niger	–	+	+	[19]
P. ostreatus	+	–	–	[19]
Cerrena maxima, Pycnoporus coccineus, Coriolopsis polyzona, P. ostreatus, T. versicolor, F. trogii	+	+	–	[4]
Trametes trogii	+	+	+	[20]
Pleurotus sajor-caju	+	+	+	[6]
Fusarium solani	+	+	+	[8]
Trametes sp. 44	+	+	+	[9]
Pleurotus eryngii (DC.) Gillet (MCC58)	+	+	+	[21]
Pleurotus ostreatus	+	+	+	[11]
Stereum ostrea	+	+	+	[22]
Dictyoarthrinium synnematicum	+	+	+	[23]
Bacteria				
Streptomyces viridosporus T7A *Streptomyces lividans* TK 64.1	–	+	+	[24]
Pseudomonas fluorescens GB-1	+	+	–	[25]
Bacillus subtilis (cot A)	+	–	–	[26]
Rhodococcus jostii	–	+	+	[27]
B. atrophaeus, B. pumilus	+	–	–	[28]

3.1.1 Laccase

Laccases (EC 1.10.3.2, benzenediol: oxygen oxidoreductase) are one of the most numerous members of the polyphenol oxidases, a blue multicopper oxidase protein family, which also includes other significant enzymes such as monooxygenases, dioxygenases and tyrosinases [29]. Laccase, a non-selective enzyme, is responsible for degradation of a wide range of substrates such as mono-, di- and poly-phenolic compounds, non-phenolic compounds and aromatic amines by oxidation to free radicals [30]. Laccases represent one of the oldest enzymes and were evidenced to be present in fungi for the first time in 1896. However, its existence was first demonstrated in a plant, the Japanese lacquer tree *Rhus vernicifera* [31]. Laccases are widely distributed in micro-organisms, insects and plants with highest yields reported in white-rot fungi, the only organisms capable of mineralizing lignin completely to carbon dioxide and water. *Coriolus hirsutus* [32], *Trametes hirsuta* [33], *T. versicolor* [4], *Pycnoporus cinnabarinus* [18] and *Pleurotus ostreatus* [11] are some of the fungus producing high quantity of laccase. Some of the bacterial laccases to be characterized are from *Azospirillum lipoferum, Bacillus subtilis*, etc. [26].

Laccase activity has also been displayed in fungal species belonging to groups other than white-rot fungi with the exception of lower fungi, Chytridiomycetes [34]. Several reports on production of laccase in ascomycetes such as *Neurospora crassa, Gaeumannomyces graminis*, *Melanocarpus albomyces*, *Magnaporthe grisea* and *Mauginella* are available in the literature [35]. In addition to these plant pathogenic species, laccase has also been reported from freshwater ascomycetes and soil ascomycete species belonging to the genera *Aspergillus, Curvularia* and *Penicillium* [36]. Furthermore, ascomycete species which participate in the decay of dead plant biomass were reported to possess laccase genes and oxidize syringaldazine, a laccase substrate [37]. Basidiomycete yeast such as *Cryptococcus neoformans* produced laccase capable of oxidizing phenols and aminophenols. However, Laccase was not found in ascomycetous yeasts, but the report of a membrane-bound multicopper oxidase enzyme Fet3p from *Saccharomyces cerevisiae* showing both structural similarity and sequence homology with laccase enzyme is available [38]. Among a number of species of white-rot fungi tested, most of them were reported to produce a varying degree of laccase enzyme [16]. Though brown-rot fungi are not known to produce laccase and no laccase protein has been purified from this group of fungi, *Coniophora puteana* was reported to oxidize syringaldazine and ABTS [39].

Fungal laccases may occur as both intracellular and extracellular isozymes with monomeric or dimeric protein structures. The significance of laccase enzyme in the process of delignification is ascribed to its catalytic efficiency to degrade the complex polyphenol structure of lignin, the most recalcitrant fraction of lignocelluloses and oxidize non-phenolic lignin model compounds mediated by certain redox agents [30]. Also, Laccase uses oxygen as a cofactor and catalyzes its reduction to water while oxidizing the phenolic substrate instead of employing expensive cofactors like biotin, pantothenic acid and thiamine [40]. Laccases exhibit a low substrate specificity and broad substrate range and can oxidize diphenols, polyphenols, methoxy-substituted phenols, aryl diamines and aminophenols and other wide range of compounds. Laccase is inducible with copper as the most commonly used inducer, and its production can also be enhanced by the presence of dyes, phenolic and non-phenolic compounds or recalcitrant compounds with structural similarity. Various properties of fungal laccases such as extracellular secretion, inducible nature, broad substrate range and use of oxygen as a cofactor have contributed to their applications in several biotechnological processes such as chemical synthesis, bioremediation of aromatic compounds, degradation of textile dyes and treatment of industrial effluents containing xenobiotic compounds [35, 41].

A typical laccase enzyme is a protein with molecular weight of 60–70 kDa that occurs in both monomeric and dimeric forms. Its primary structure in the form of Greek key β barrel topology is composed of approximately 500 amino acid residues distributed in three successive domains with 150 initial amino acids in the first domain, 150–300 in the second domain and 300–500 amino acid in the third domain. Two disulphide bridges located between domains I and II and between I and III are responsible for stabilization of the enzyme structure [42]. However, presence of a third disulphide bridge inside domain I has also been demonstrated in some laccases. The catalysis leading to oxidation of substrates along with reduction of molecular oxygen to water is due to the presence of active centres containing four copper (Cu) molecules in the enzyme [43]. The enzyme contains these Cu molecules in different cores comprising one type 1 Cu (T1), two type 2 Cu (T2) and one type 3 Cu (T3) resulting in the formation of two metallic active sites: a mononuclear centre of one T1 and a trinuclear cluster containing two T2 and one T3 [42]. T1 has the highest redox potential and is responsible for oxidizing the substrate while T2/T3 core receives the electrons and carries out the reduction of oxygen to water. It also interacts with inhibitory ions like fluoride and assists the inhibition of the enzymatic activity [35]. Several isozymes of laccase possessing different molecular weight and isoelectric point have been reported from white-rot fungi such as *P. ostreatus* and *T. versicolor*. A significant structural characteristic of these isozymes is the presence of different types of ligands such as ions (Ca^{2+}), monosaccharides (mannose, N-acetyl glucosamine) and organic molecules [40, 42]. An example of the presence of these ligands in fungal strains is the glycosylations exhibited in the laccase isoenzymes of *Trametes versicolor* [44].

3.1.2 Heme-Peroxidases

Lignin peroxidase (LiP) and manganese peroxidase (MnP), two most significant ligninolytic enzymes after laccase, belong to the class II family of heme-peroxidases superfamily. These enzymes belong to this superfamily based on their catalytic properties and the presence of protoporphyrin IX as a prosthetic group. Similar to horseradish peroxidase, a well-known heme-peroxidase the catalytic cycle of LiP and MnP to oxidize their respective substrates involves multi-step electron transfer reactions with the formation of intermediate radical cation [43]. H_2O_2 oxidizes the enzyme to produce a modified enzyme (Compound I) and water followed by the production of a free radical and another modified form of the enzyme (Compound II) mediated by an electron transfer from the substrate (SH: reduced substrate). Finally, this Compound II reacts with reduced form of substrate producing another free radical and water.

3.1.2.1 Lignin Peroxidase

Lignin peroxidases [EC 1.11.1.14, 1,2-bis(3,4-dimethoxyphenyl)propane-1,3-diol: hydrogen peroxide oxidoreductase], the enzymes responsible for H_2O_2-dependent oxidative depolymerization of lignin, were originally discovered in nitrogen and carbon limited cultures of *Phanerochaete chrysosporium* [43, 45]. LiP is relatively non-specific in nature and has the tendency to oxidize phenolic substrates, non-phenolic lignin model compounds and a variety of organic compounds [45]. It possesses high redox potential of 700–1400 mV, a low optimum pH of 3–4.5 and the ability to degrade a wide number of aromatic compounds such as veratryl alcohol and methoxybenzenes [43]. LiP catalyzes the cleavage of C_α–C_β bonds and β-o-4 ether bonds of lignin, thus causing the depolymerization of lignin [46]. It also catalyzes hydroxylation reaction, aromatic ring cleavage and quinine formation. The basic catalytic cycle of LiP involves formation of three different forms of the enzymes (Compound I, II and III) and veratryl alcohol radicals with high redox potential. These VA radicals can participate in the degradation of highly activated aromatic substrates with high redox potential [44, 47]. Oxidation of phenolic compounds such as guaiacol and syringic acid occurs at a much faster compared to non-phenolic compounds and is mediated by the formation of phenoxy radicals which in the presence of O_2 may react to form ring-cleavage products or may also cause coupling and polymerization.

LiPs are monomeric glycoproteins of 38–46 kDa and are composed of approximately 343 amino acids, a carbohydrate chain, two calcium-binding sites and a heme group [44, 45 and 47]. It is a globular protein composed of a C-terminal and N-terminal domain with the heme group embedded between them. The secondary structure of the enzyme molecule contains eight major helices, eight minor helices and two anti-parallel beta sheets in the C-terminal domain. The calcium-binding sites are possibly responsible for maintaining the topology of the active site. The heminic cavity consists of 40 amino acid residues, and the heme iron coordinates with His176 at proximal side and is hydrogen bonded to His46 at the distal side. This "His" in its reduced imidazole character and its distance from the heme group is responsible for the high redox potential of LiP [44, 47]. The presence of a tryptophan residue (Trp171) on the enzymes' surface further increases the redox potential of LiP. Also, Trp171 residue plays a significant role in binding and catalysis of veratryl alcohol by electron transfer to the enzyme which in turn participates in the oxidation of different aromatic molecules [45]. Besides acting as a redox mediator, veratryl alcohol also protects the enzyme from the action of H_2O_2 [47]. The LiP activity is easily measured spectrophotometrically as the increase in absorbance per minute due to oxidation of veratryl alcohol to veratraldehyde (VAD) at 310 nm.

Various isoforms of LiP are known to exist with *Phanerochaete chrysosporium*, and the molecular structure of *P. chrysosporium* LiP from has been extensively studied. Apart from five different strains of *Phanerochaete*, the LiP production has also observed in other white-rot fungi namely *Coriolus versicolor*, *Lentinula edodes*, *P. magnoliae*, *Phellinus pini*, *P. radiata*, *Trametes hirsute*, *Trametes gibbosa* and *Trametes trogii* [48].

3.1.2.2 Manganese Peroxidase

MnP [EC 1.11.1.13, Mn(II):hydrogen peroxide oxidoreductase, MnP] is a Mn(II) and H_2O_2-dependent enzyme which oxidizes a wide range of substrates including lignin, phenols, phenolic lignin model compounds, dyes, amines, high molecular weight chlorolignins and non-phenolic compounds [46]. However, its significance in delignification is mainly for its ability to attack and degrade phenolic lignin component. It catalyzes the Mn-dependent reaction by oxidizing Mn^{2+} to Mn^{3+} in the presence of H_2O_2. The formed Mn^{3+}, a strong oxidant, then oxidizes phenolic structures by single-electron oxidation [44, 45]. The extracellular MnP was first reported and purified from culture of *P. chrysosporium,* and its production is limited to certain white-rot fungi producing multiple isozymes of MnP.

It is a heme-containing acidic glycoprotein with molecular weight of 38-62.5 kDa, composed of 350 amino acid residues and structural and sequence similarity with LiP [49]. The protein molecule contains two domains with the heme group in the middle, two calcium ions, ten major helices, a minor helix and five disulphide bonds. One additional disulphide bond, otherwise not present in other peroxidases, participates in the formation of manganese (Mn) binding site. The enzyme follows the usual peroxidase catalytic cycle, but with Mn(II) as the reducing substrate. MnP catalyzes the oxidation of Mn(II) to Mn(III), which in turn mediates the oxidation of a variety of organic substrates. Organic acid chelators such as oxalic acid are significant to the catalytic cycle since MnP reacts with chelated and not free Mn(II), and they also stabilize and dissociate Mn(III) from the enzyme [49, 50]. Mn(III) is a strong oxidizing agent that oxidizes a variety of phenolic compounds, but cannot oxidize non-phenolic substrates by itself and requires a second mediator [50]. The catalytic cycle begins with the $2e^-$ transfer from the heme group of native MnP to H_2O_2, thus yielding MnP-I, a Fe(IV)-oxo-porphyrin radical cation and water. MnP-I then catalyzes the oxidation of Mn^{2+} accompanied by reduction of MnP-I to MnP-II. MnP-II oxidizes Mn^{2+} to produce Mn^{3+}, the cation responsible to oxidize aromatic compounds. The conversion of MnP-I to MnP-II can also be mediated by electron donors such as ferrocyanide instead of Mn^{2+}, but the reduction of MnP-II to its native form requires Mn^{2+} as a redox coupler for completion of its catalytic cycle. After Mn^{3+} is dissociated from the enzyme and stabilized by chelating compounds, it diffuses in and attacks the lignified plant cell wall which facilitates the penetration of other enzymes [45, 49].

3.2 Hemicellulases

Hemicellulases are mostly modular proteins containing structurally discrete catalytic and non-catalytic modules. Non-catalytic modules of hemicellulase include carbohydrate-binding modules which accelerate the targeting of the hemicellulase enzyme to the polysaccharide and dockerin modules which are responsible for the binding of catalytic module to the microbial cell surface or multi-enzyme complexes such as cellulosomes [51]. On the basis of amino acid sequences of catalytic modules, hemicellulases are characterized as glycoside hydrolases which hydrolyze glycosidic bonds of hemicellulose and esterases which cleave the ester linkages of acetate or ferulic acid side groups. Hemicellulases have also been grouped into various families based on their primary sequence homology.

3.2.1 Xylanases and Mannanases

Due to the complex branched structure and heterogeneity of hemicelluloses, synergistic action of several different enzymes is required for complete hydrolysis of hemicellulose to its constituent sugar monomers. Out of these, the two main hydrolases are endo-1,4-β-D-xylanase and endo-1,4-β-D-mannanase which depolymerize the hemicellulose backbone [52]. Endo-1,4-β-xylanase is responsible for cleaving the glycosidic bonds in the xylan backbone, thus causing a reduction in the degree of polymerization of the substrate. Enzymatic attack on xylan is not random, and the bonds selected are dependent on the length of backbone chain, the degree of branching and substituents in the substrate molecule [53]. β-Dxylopyranosyl oligomers are the initial hydrolysis product; however, smaller molecules such as mono-, di- and trisaccharides may also be produced during later stages of hydrolysis [54]. Hemicellulases are produced by both bacteria and fungi. Filamentous fungi are the main producers of extracellular xylanases with enzyme yields obtained much higher than yeasts and bacteria [55]. *Aspergillus niger*, *Humicola insolens*, *Trichoderma reesei*, *Thermomonospora fusca*, *T. longibrachiatum* and *T. koningii* have been used in industry for commercial production of xylanases. Among Bacteria, xylanases are mostly produced by *Bacillus* sp. Xylanases have many industrial applications such as biobleaching of wood pulps in paper industry, juice and wine industries, baking industry, ingredients in detergents or fabric care solutions and xylitol production [54, 56]. Xylanases are usually inducible enzymes, and the biosynthesis can be induced by the addition of xylan, xylose and xylobiose to the culture medium [54, 55 and 56]. Xylan is a complex but major type of hemicelluloses in wood, and the complete hydrolysis of xylan to xylose requires the cooperative action of a multi-enzymatic system composed of β-xylanase and β-xylosidase and other accessory enzymes such as acetylxylan esterase, α-L-arabinofuranosidase, α-glucuronidase and hydroxycinnamic acid esterases that are responsible for cleavage of side chain residues from the xylan backbone [52]. Xylanases attack the backbone of xylan to produce shorter chain oligomers, xylobiose and xylose. Xylosidases are essential for the hydrolysis of xylooligosaccharides to xylose [57]. The enzymes arabinosidase, α-glucuronidase and acetylxylan esterase are required to release the substituents on the xylan backbone to achieve complete hydrolysis of xylan to monosaccharides [58].

Galactoglucomannan is mainly composed of D-mannose. However, complete breakdown of Galactoglucomannan into simple sugars requires the cooperative action of endo-1,4-β-mannanases (EC 3.2.1.78) and exoacting β-mannosidases (EC 3.2.1.25) to cleave hemicellulose backbone and accessory enzymes, such as β-glucosidases (EC 3.2.1.21), α-galactosidases (EC 3.2.1.22) and acetyl mannan esterases to further remove the side chain sugar residues [59]. A vast variety of bacteria, actinomycetes, yeasts and fungi are known to produce these enzymes. These enzymes are usually extracellular and have been reported to be both inducive and constitutive [59]. Although a large number of bacterial species are known to produce mannanases, only a few commercially exploited ones include *Bacillus* sp., *Caldibacillus cellulovorans*, *Caldocellum saccharolyticum* and *Streptomyces* sp. [57, 60].

3.2.2 Xyloglucanases

According to a new classification of glycoside hydrolases available at http://afmb.cnrs-mrs.fr/CAZY/, some endo-β-1,4-glucanases with high affinity towards xyloglucan were assigned a new EC number (EC 3.2.1.151) and designated as xyloglucanase, xyloglucan hydrolase (XGH) or xyloglucan-specific endo-β-1,4-glucanases (XEGs) [61]. These enzymes represent a new class of hemicellulases with the ability to attack the substituted glucose residues in the backbone. In addition to these enzymes, β-galactosidase which cleaves the galactose residue attached to the branched xylose moiety in the β-D-glucopyranose backbone is also required for complete breakdown of galactoxyloglucan [62]. Among these, xyloglucanases are found as both endo-type and exo-type hydrolases in several micro-organisms and have the highest specific activity towards xyloglucan [63–65]. The endo-type enzymes catalyze random hydrolysis of xyloglucan polymer while the exo-type enzymes recognize specific ends of xyloglucan. Some examples of exo-type hydrolases are oligoxyloglucan reducing end-specific cellobiohydrolase from *Geotrichum* sp. M128; xyloglucanobiohydrolase from *Aspergillus nidulans* [65]; XEG74 from *Paenibacillus* sp. KM21; and Cel74A from *Trichoderma reesei* [63, 64].

3.3 Cellulases

Cellulases are glycoside hydrolases (GHs) that degrade cellulose by hydrolysis of β-1,4-glycosidic bonds of this hydrophilic, water-insoluble polymer into shorter chain polysaccharides such as cellodextrans, cellobiose and glucose. Cellulases typically have two separate domains: a catalytic domain that hydrolyzes the glycosidic bonds and a cellulose-binding module that facilitates the transfer of catalytic domain to the cellulose molecule. They are linked together by a flexible linker region which is rich in serine and threonine residues. Some cellulases also possess ancillary modules such as FN3-like modules [66].

The cellulase system consists of three major enzyme components: β-endoglucanase (endo-1,4-b-D-glucanase, EC 3.2.1.4), β-exoglucanase (1,4-b-D-glucan-cellobiohydrolase, EC 3.2.1.91) or cellobiohydrolases and β-D-glucosidase (β-Dglucosideglucanohydrolase, EC 3.2.1.21) or cellobiases that work collaboratively for complete hydrolysis of the complex crystalline cellulose microfibrils to glucose [67]. Endoglucanases breakdown the internal bonds of the glycan chains randomly causing deconstruction of the amorphous fraction of the cellulose producing mainly cellodextrin and also provide reducing or non-reducing ends of cellooligosaccharides for cellobiohydrolases. Cellobiohydrolases then target the crystalline cellulose either at the reducing end or at the non-reducing end, yielding cellobiose. Lastly, β-glucosidases catalyze the cleavage of cellobiose to produce glucose and also release glucose from the non-reducing ends of soluble cellooligosaccharides [68, 69]. Thus, β-glucosidase plays an important role by converting cellobiose to glucose since it is the end-product inhibitor of both exo- and endoglucanases [69].

The endo- and exoglucanases differ structurally in terms of the site (loop) for cellulose binding [70]. Endoglucanases are characterized by the presence of small loops with open active binding site for any accessible region along the cellulose chains to produce oligomers of different lengths [70, 71]. On the other hand, exoglucanases consist of long loops with affinity for the crystalline regions and produce cellodextrin as the major hydrolysis product [71]. Also, exoglucanases can be present as either the reducing or non-reducing end cellobiohydrolases or both but act uni-directionally to assure the complete hydrolysis of the polysaccharide [70].

Cellulases have been sourced from different organisms including fungi, bacteria and protozoans, but fungi are the most studied cellulose-degrading micro-organisms owing to their high enzymes secretion capabilities [70]. The most studied cellulolytic enzymes are from *T. reesei* which are extensively used in commercial cellulase preparations [68]. The cellulase mixtures obtained from *T. reesei* consist of almost 80% exoglucanases, approximately 15% endoglucanases and lesser amounts of other accessory enzymes [66]. *T. reesei* also produces intracellular β-glucosidase, but the levels are usually too low to fight cellobiose feedback inhibition during cellulose hydrolysis and may require boost from other sources [72]. Other examples of the most significant cellulase producing organism are filamentous fungi of the genus *Aspergillus* and include *A. niger*, *Aspergillus oryzae*, *A. fumigatus*, *A. unguis* and *A. tubingensis* [73]. Other fungi reported to produce cellulases include *Sclerotium rolfsii* and species of *Schizophyllum, Fusarium* and *Penicillium*. Although most of the cellulases used in enzymatic hydrolysis of lignocellulosic biomass are those derived from fungi, yet the most efficient natural cellulolytic system known is actually produced by bacterial system. This is attributed to the efficient heterologous production and less rigorous pH requirement of bacterial systems and high specific activity of enzymes produced from bacteria [74]. *Bacillus*, *Cellulomonas*, *Streptomyces*, *Cytophaga*, *Cellvibrio* and *Pseudomonas* are some of the bacterial genera known for their cellulolytic potential. Among these, *B. subtilis* has been exploited extensively for its high protein secretion capacity and high-activity endoglucanase in metabolic engineering to develop recombinant strains capable of growth on cellulose as a sole carbon source [75].

References

1. Costa SM, Goncalves AR, Esposito E (2005) *Ceriporiopsis Subvermispota* used in delignification of sugarcane bagasse prior to soda/anthraquinone pulping. In: Twenty-sixth symposium on biotechnology for fuels and chemicals. Humana Press, 695–706
2. Hao JJ, Tian XJ, Song FQ, He XB, Zhang ZJ, Zhang P (2006) Involvement of lignocellulolytic enzymes in the decomposition of leaf litter in a subtropical forest. J Eukar Microbiol 53:193–198
3. Levin L, Herrmann C, Papinutti VL (2008) Optimization of lignocellulolytic enzyme production by the white-rot fungus *Trametes trogii* in solid-state fermentation using response surface methodology. Biochem Eng J 39(1):207–214
4. Elisashvili V et al (2008) *Lentinus edodes* and *Pleurotus* species lignocellulolytic enzymes activity in submerged and solid-state fermentation of lignocellulosic wastes of different composition. Bioresour Technol 99:457–462

5. Elisashvili V, Kachlishvili E, Khardziani T, Agathos SN Effect of aromatic compounds on the production of laccase and manganese peroxidase by white-rot basidiomycetes. J Ind Microbiol Biotechnol 37: 1091–1096
6. Massadeh MI, Fraij A, Fandi K (2010) Effect of carbon sources on the extracellular lignocellulolytic enzymetic system of *Pleurotus Sajor-Caju*. Jordan J Biol Sci 3(2):51–54
7. Hu HL, Vanden Brink J, Gruben BS, Wosten HAB, Gu JD, Deveries RP (2011) Improved enzyme production by co-cultivation of *A. niger* and *A. oryzae* and with other fungi. Inter Biodeterior Biodegradation 65:248–252
8. Obruca S, Marova I, Matouskova P, Haronikova A, Lichnova A (2012) Production of lignocellulose-degrading enzymes employing *Fusarium solani* F-552. Folia Microbiol 57:221–227
9. Gomez SQ, Arana-Cuenca A, Flores YM, Rodriguez JNG, Tellez-Jurado A (2012) Effect of particle size and aeration on the biological delignification of corn straw using *Trametes* sp. 44. Bioresources 7(1):327–344
10. Faten AM, El Aty Abd, Abeer A (2013) Enzyme activities of the marine-derived fungus *Alternaria alternata* cultivated on selected agricultural wastes. J App Bio Sci 7(1):39–46
11. Kheiralla ZH, Badr El-Din SM, Abdel Malek SM, Abdel Aziz DH (2013) Optimization of cultural conditions for lignin peroxidase production by *Phanerochaete chrysosporium* and *Pleurotus ostreatus*. Academia J Biotechnol 1(6):087–095
12. Ma K, Ruan Z (2015) Production of a lignocellulolytic enzyme system for simultaneous bio-delignification and saccharification of corn stover employing co-culture of fungi. Bioresour Technol 175:586–593
13. Meehnian H, Jana AK, Jana MM (2016) Effect of particle size, moisture content, and supplements on selective pretreatment of cotton stalks by *Daedalea flavida* and enzymatic saccharification. 3 Biotech 6: 235
14. Bonnarme P, Asther M (1993) Influence of primary and secondary proteases produced by free or immobilized cells of the white-rot fungus *Phanerochaete chrysosporium* on lignin peroxidase activity. J Biotechnol 30:271–282
15. Linares NC, Ortiz DM, Doralinda A, Guzman-Ortiz DA, Fernandez F, Loske AM, Gomez-Lim MA (2014) High-yield production of manganese peroxidase, lignin peroxidase, and versatile peroxidase in *Phanerochaete chrysosporium*. App Microbiol Biotechnol 98:9283–9294
16. Hatakka A and Niemenmaa O (1991) Production and purification of peroxidases from the white rot fungi *Phlebia radiata* and *Phlebia tremallosa*. In: Finnish-Soviet seminar on bioconversion on plant raw materials—biotechnology advancement, pp 189–204
17. Galliano H, Gas G, Series JL, Boudet AM (1991) Lignin degradation by *Rigidoporus lignosus* involves synergistic action of two oxidizing enzymes Mn peroxidase and laccase. Enzyme Microb Technol 13(6):478–482
18. Saparrat MCN, Martinez MJ, Tournier HA, Cabello MN, Arambarri AM (2002) Screening for ligninolytic enzymes in autochthonous fungal strains from Argentina isolated from different substrata. Revista Iberoamericana De Micologia 19:181–185
19. Pant D, Adholeya A (2007) Enhanced production of ligninolytic enzymes and decolorization of molasses distillery wastewater by fungi under solid state fermentation. Biodegradation 18:647–659
20. Patrick F, Mtui G, Mshandete AM, Kivaisi A (2010) Optimized production of lignin peroxidase, manganese peroxidase and laccase in submerged cultures of *Trametes trogii* using various growth media compositions. Tanz J Sci 36
21. Akpinar M, Urek RO (2012) Production of ligninolytic enzymes by solid state fermentation using *Pleurotus eryngii*. Preparative Biochem Biotechnol 42:582–597
22. Usha KY, Praveen K, Reddy BR (2014) Enhanced production of ligninolytic enzymes by a Mushroom *Stereum ostrea*. Biotechnol Res Int 815495
23. Prasher IB, Chauhan R (2015) Effect of carbon and nitrogen sources on the growth, reproduction and ligninolytic enzymes activity of *Dictyoarthrinium synnematicum*. Adv Zoo Botany 3(2):24–30

24. Korus RA, Lodha SJ, Adhi TP, Crawford DL (1991) Kinetics of peroxidase production by *Streptomyces viridosporus* and recombinant *Streptomyces lividans*. Biotechnol Prog 7:510–515
25. Okazaki M, Sugita T, Shimizu M, Ohode Y, Iwamoto K, Jong V, de Vrind JPM, Corstjens PLAM (1997) Partial purification and characterization of mangenese-oxidizing factors of *Pseudomonas fluorescens* GB-1. App Environ Microbiol 63:4793–4799
26. Hullo MF, Moszer I, Danchin A, Martin-Verstraete I (2001) CotA of *Bacillus substilis* is a copper-dependent laccase. J Bacteriol 183:5426–5430
27. Ahmad M, Roberts JN, Hardiman EM, Singh R, Eltis LD, Bugg TDH (2011) Identification of DypB from *Rhodococcus jostii* RHA1 as a lignin peroxidase. Biochemistry 50:5096–5107
28. Huang XF, Santhanam N, Badri DV, Hunter WJ, Manter DK, Decker SR, Vivanco JM, Reardon KF (2013) Isolation and characterization of lignin-degrading bacteria from rainforest soils. Biotechnol Bioeng 110:1616–1626
29. Sharma RK, Arora DS (2010) Production of lignocellulosic enzymes and enhancement of invitro digestibility during solid state fermentation of wheat straw by *Phlebia floridensis*. Bioresour Technol 101:9248–9253
30. Leonowicz A, Cho NS, Luterek J, Wilkolazka A, Wotjas-Wasilewska M, Matuszewska A, Hofrichter M, Wesenberg D, Rogalski J (2001) Fungal laccase: properties and activity on lignin. J Basic Microbiol 41:185–227
31. Yoshida H (1983) Chemistry of lacquer (urushi). J Chem Soc 43:472–486
32. Koroleva OV, Gavrilova VP, Stepanova EV, Lebedeva VI, Sverdlova NI, Landesman EO (2002) Production of lignin modifying enzymes by co-cultivated white rot fungi *Cerrena maxima* and *Coriolus hirsutus* and characterization of laccase from *Cerrena maxima*. Enzyme Microb Technol 30:573–580
33. Rodríguez-Couto S, Rodríguez A, Paterson RRM, Lima N, Teixeira JA (2006) High laccase activity in a 6 l airlift bioreactor by free cells of *Trametes hirsuta*. Lett Appl Microbiol 42:612–616
34. Morozova OV, Shumakovich GP, Gorbacheva MA, Shleev SV, Yaropolov AI (2007) "Blue" laccases. Biochemistry 72(10):1136–1150
35. Baldrin P (2006) Fungal laccases—occurrence and properties. FEMS Microbiol Rev 30:215–242
36. Junghanns C, Moeder M, Krauss G, Martin C, Schlosser D (2005) Degradation of the xenoestrogen nonylphenol by aquatic fungi and their laccases. Microbiology 151(1):45–57
37. Lyons JI, Newell SY, Buchan A, Moran MA (2003) Diversity of ascomycete laccase gene sequences in a southeastern US salt marsh. Microb Eco 45(3):270–281
38. Stoj C, Kosman DJ (2003) Cuprous oxidase activity of yeast Fet3p and human ceruloplasmin: implication for function. FEBS Lett 554(3):422–426
39. Lee KH, Wi SG, Singh AP, Kim YS (2004) Micromorphological characteristics of decayed wood and laccase produced by the brown-rot fungus *Coniophora puteana*. J Wood Sci 50(3):281–284
40. Matera I, Gullotto A, Tilli S, Ferraroni M, Scozzafava A, Briganti F (2008) Crystal structure of the blue multicopper oxidase from the white-rot fungus *Trametes trogii* complexed with p-toluate. Inorg Chim Acta 361:4129–4137
41. Chanaga Vera X, Placido Escobar J, Marin Montoya M, Perez Y, Del Socorro M (2012) Native fungi with industrial dye degrading potential in the Aburra Valley, Colombia. Revista Facultad Nacional de Agronomia, Medellin 65:6811–6821
42. Ferraroni M, Myasoedova N, Schmatchenko V, Leontievsky A, Golovleva L, Scozzafava A, Briganti F (2007) Crystal structure of a blue laccase from *Lentinus tigrinus*: evidences for intermediates in the molecular oxygen reductive splitting by multicopper oxidases. BMC Struct Biol 7:60
43. Dias A, Sampaio A, Bezerra R (2007) Environmental applications of fungal and plant systems: decolourisation of textile wastewater and related dyestuffs. In: Environmental bioremediation technologies. Springer, Berlin, Heidelberg, 445–463
44. Wong DWS (2009) Structure and action mechanism of ligninolytic enzymes. Appl Biochem Biotechnol 157:174–209

45. Hammel KE, Cullen D (2008) Role of fungal peroxidases in biological ligninolysis. Curr Opin Plant Biol 11:349–355
46. Eriksson KEL (2000) Lignocellulose, lignin, ligninases. In: Encyclopedia of microbiology. Academic press, San Diego, 3, 39–48
47. Choinowski T, Blodig W, Winterhalter KH, Piontek K (1999) The crystal structure of lignin peroxidase at 1.70 Å resolution reveals a hydroxy group on the Cβ of tryptophan 171: a novel radical site formed during the redox cycle. J Mol Biol 286:809–827
48. Ahammed S, Prema P (2002) Influence of media nutrients on synthesis of lignin peroxidase from *Aspergillus* sp. Appl Biochem Biotechnol 103:327–336
49. Martin H (2002) Review: lignin conversion by manganese peroxidase (MnP). Enzyme Microb Technol 30:454–466
50. Gold MH, Youngs HL, Sollewijn Gelpke MD (2000) Manganese peroxidase. Metal ions biological systems. Marcel Dekker, New York, pp 559–587
51. Shallom D, Shoham Y (2003) Microbial hemicellulases. Curr Opin Microbiol 6:219–228
52. Biely P (1993) Biochemical aspects of the production of microbial hemicellulases. In: Hemicellulose and Hemicellulases. Portland Press, Cambridge, 29–51
53. Li KC, Azadi P, Collins R, Tolan J, Kim JS, Eriksson KEL (2000) Relationships between activities of xylanases and xylan structures. Enzyme Microb Technol 27:89–94
54. Polizeli MLTM, Rizzatti ACS, Monti R, Terenzi HF, Jorge JA, Amorim DS (2005) Xylanases from fungi: properties and industrial applications. Appl Microbiol Biotechnol 67:577–591
55. Kulkarni N, Shendye A, Rao M (1999) Molecular and Bioechnological aspects of xylanases. FEMS Microbiol reviews 23:411–456
56. Beg QK, Kapoor M, Mahajan L, Hoondal GS (2001) Microbial xylanases and their industrial applications: a review. Appl Microbiol Biotechnol 56:326–338
57. Deshpande V, Lachke A, Mishra C, Keskar S, Rao M (1986) Mode of action and properties of xylanase and L-xylosidase from Neurospora crassa. Biotechnol Bioeng 26:1832–1837
58. Gilbert HJ, Hazlewood GP (1993) Bacterial cellulases and xylanases. J Gen Microbiol 139:187–194
59. Dhawan S, Kaur J (2007) Microbial mannanases: an overview of production and applications. Crit Rev Biotechnol 27:197–216
60. Morris DD, Reeves RA, Gibbs MD, Saul DJ, Bergquist PL (1995) Correction of the β-mannanase domain of the cell C pseudogene from Caldocellulosiruptor saccharolyticus and activity of the gene product on kraft pulp. Appl Environ Microbiol 61:2262–2269
61. Henrissat B, Bairoch A (1996) Updating the sequencebased classification of glycosyl hydrolases. Biochem J 316:695–696
62. Benko Z, Siikaaho M, Viikari L, Reczey K (2008) Evaluation of the role of xyloglucanase in the enzymatic hydrolysis of lignocellulosic substrates. Enzyme Microb Technol 43:109–114
63. Yaoi K, Nakai T, Kameda Y, Hiyoshi A, Mitsuishi Y (2005) Cloning and characterization of two xyloglucanases from *Paenibacillus* sp. strain KM21. Appl Environ Microbiol 71:7670–7678
64. Grishutin SG, Gusakov AV, Markov AV, Ustinov BB, Semenova MV, Sinitsyn AP (2004) Specific xyloglucanases as a new class of polysaccharide-degrading enzymes. Biochim Biophys Acta 1674:268–281
65. Bauer S, Vasu P, Mort AJ, Somerville CR (2005) Cloning, expression, and characterization of an oligoxyloglucan reducing end-specific xyloglucanobiohydrolase from *Aspergillus nidulans*. Carbohydr Res 340:2590–2597
66. Garvey M, Klose H, Fischer R et al (2013) Cellulases for biomass degradation: comparing recombinant cellulase expression platforms. Trends Biotechnol 31:581–593
67. Lynd LR, Weimer PJ, Van Zyl WH, Pretorius IS (2002) Microbial cellulose utilization: fundamentals and biotechnology and microbiology. Mol Biol Rev 66:506–577
68. Wahlstrom R, Rahikainen J, Kruus K, Suurnakki A (2014) Cellulose hydrolysis and binding with *Trichoderma reesei* Cel5A and Cel7A and their core domains in ionic liquid solutions. Biotechnol Bioeng 111:726–733
69. Jorgensen H, Kristensen JB, Felby C (2007) Biofuels. Bioprod Biorefin 1:119–134

70. Juturu V, Wu JC (2014) Microbial cellulases: engineering, production and applications. Renew Sustain Energy Rev 33:188–2037
71. Segato F, Damásio ARL, de Lucas RC et al (2014) Genome analyses highlight the different biological roles of cellulases. Microbiol Mol Biol Rev 78:588–613
72. Parisutham V, Kim TH, Lee SK (2014) Feasibilities of consolidated bioprocessing microbes: from pretreatment to biofuel production. Bioresour Technol 161:431–440
73. Obeng EM, Adam SNN, Budiman C, Ongkudon CM, Maas R, Jose J (2017) Lignocellulases: a review of emerging and developing enzymes, systems, and practices. Bioresour Bioprocess 4:16
74. Stern J, Kahn A, Vazana Y et al (2015) Significance of relative position of cellulases in designer cellulosomes for optimized cellulolysis. PLoS ONE 10:e0127326. https://doi.org/10.1371/journal.pone.0127326
75. Zhang YHP (2011) Substrate channeling and enzyme complexes for biotechnological applications. Biotechnol Adv 29:715–725

Chapter 4
Pretreatment Strategies: Unlocking of Lignocellulosic Substrate

Abstract Conversion of lignocelluloses to bioethanol takes place in three main stages, namely pretreatment to remove lignin and expose the crystalline structure of cellulose; enzymatic hydrolysis to convert cellulose and hemicellulose to simple sugars and microbial fermentation of sugars to ethanol. The most important of these steps is the pretreatment step employed to eliminate lignin and reduce the crystallinity of cellulose to make it accessible for enzymatic hydrolysis for its conversion to glucose. The pretreatment step controls the efficiency of subsequent steps and also accounts for the maximum part of the production cost of biofuel from lignocellulosics. A large number of pretreatment methods including physical methods, chemical treatment, physico-chemical processes, thermo-chemical pretreatment and biological pretreatment are available. Among the various pretreatment methods, biological pretreatment is a promising approach because of low operational cost, does not produce waste and other compounds (phenolics) toxic to the fermenting micro-organisms and is less energy-intensive and environment-friendly. This chapter discusses some of the important pretreatment methods for the disruption of complex lignocellulosic structure and production of fermentable sugars.

Keywords Biological · Cellulose · Hydrolysis · Lignin · Ozonolysis · Pretreatment

4.1 Pretreatment Technologies for Lignocellulosic Biomass

Lignocellulosic feedstocks such as forest, agricultural, industrial and agroindustrial residues are among the most sustainable sources of biomass for bioethanol production and other biochemicals. However, lignocelluloses as a substrate for bioethanol production come with a certain set of challenges not associated with molasses or starch-based biofuel generation [1]. First, an efficient pretreatment step is required to disrupt the complex structure of lignocellulose, increase the surface area, decrease the degree of polymerization and crystallinity and expose cellulose and hemicellulose to enzymatic hydrolysis. Secondly, the exposed cellulose and hemicellulose fraction must be broken down into simpler sugars during enzymatic hydrolysis in an efficient and cost-effective manner. Finally, these sugars as a complex mixture of

A. Sharma and N. K. Aggarwal, *Water Hyacinth: A Potential Lignocellulosic Biomass for Bioethanol*, https://doi.org/10.1007/978-3-030-35632-3_4

pentose's and hexoses must be fermented to bioethanol. Low conversion efficiency due to the natural recalcitrance of lignin to deconstruction and high cost associated with enzymatic conversion are some of the major challenges [2]. An effective pretreatment technique is, therefore, essential to prepare cellulose for enzymatic saccharification step with high yield [3]. An "ideal" pretreatment method should be simple and economical, require less energy, process water and chemicals, efficiently disrupt the lignocellulosic structure, must be selective in nature to minimize loss of sugars, results in the limited generation of inhibitory products like phenolic acids, furfural or 5-hydroxymethylfurfural [4]. A wide range of physical, chemical, physico-chemical, thermo-chemical and biological pretreatment methods and combinations thereof have been reported for degradation and enhanced enzymatic saccharification of lignocellulosic biomass. A list of various pretreatment methods along with their advantages and disadvantages is given in Table 4.1.

4.1.1 Physical Methods

This method is solely focused on the destruction of the lignin structure, reduction in particle size in order to increase the surface area and decrease in the degree of polymerization and crystallinity and thus is a prerequisite before any other pretreatment method. Physical pretreatment includes various mechanical and non-mechanical methods. Mechanical methods like chipping, grinding and milling are significant in improving the bioconversion effectivity, particle densification, enzymes accessibility and thus overall transformation of lignocellulosic biomass into biofuels [5]. Though milling is known to reduce the particle size of material up to 0.2 mm, decrease in particle size and degree of crystallinity is also dependent on the type of biomass and mill used and the processing time. Depending upon the type of milling equipment used, the different milling methods include centrifugal milling, colloid milling, hammer milling, two-roll milling, ball milling, rod milling, colloid milling, wet disc milling and vibratory milling [6, 7]. Extrusion is another physical method in which the substrate is subjected to heating, mixing and shearing leading to both physical and chemical reconstruction [8]. It is a promising thermo-mechanical pretreatment method due to its continuous nature, high versatility, efficient mixing and heat transfer capabilities and operation at high loadings.

Non-mechanical method involves microwave treatment and ultrasonication. Microwave irradiation is a non-conventional pretreatment technique which employs an applied electromagnetic field for disruption of complex lignocellulosic structure. This disruption is caused by thermal energy generated on molecular collisions due to dielectric polarization [9]. Microwave irradiation can also be operated at high pressure in closed reactors at a very high temperature ranging from 150 to 250 °C [10]. Microwave pretreatment is also often integrated with other treatment methods to increase the efficiency of pretreatment. During a study on the impact of microwave irradiation on alkaline pretreatment, it was discovered that microwave application effectively fractured the complex structure of hardwood promoting the penetration of

Table 4.1 Advantages and disadvantages of different pretreatment methods of lignocellulosic biomass

Pretreatment method	Advantages	Disadvantages
Physical methods (chipping, grinding and milling)	Reduce cellulose crystallinity Simple method	High energy consumption High cost of equipment
Alkali	Efficient lignin removal Reduced inhibitors production	High cost of catalyst Alteration of lignin structure
Acid	High glucose yield Hemicellulose solubilization	High cost of acids, recovery and corrosive resistant equipment Formation of inhibitors
Steam explosion	Cost-effective Lignin degradation and hemicellulose solubilization High yield of glucose	Partial degradation of hemicellulose Acid catalyst required for high lignin content material Toxic compounds generation
LHW	Separation of hemicellulose from rest of feedstock and improved hydrolysis No catalysts needed	High energy/water input Leftover solid mass (cellulose/lignin) needed to be dealt with
AFEX	Highly effective method for herbaceous and low lignin content biomass Increased cellulose accessibility Low inhibitors formation	High cost of ammonia Recycling of ammonia needed Low effectiveness with increased lignin content Alters lignin structure
CO_2 explosion	Cost-effective Does not form inhibitory compounds	Low yield Very high pressure requirement
Biological pretreatment	Cost-effective Mild Eco-friendly	Slow hydrolysis rate Degradation of hemicellulose and cellulose along with lignin

alkaline solution into the inner fibre structure resulting in significant hemicelluloses disruption and high cellulose yield [11]. Various advantages of microwave treatment include easy operation, high heating capacity in a short span of time, energy efficiency and minimum inhibitors formation [12]. Ultrasonication pretreatment is based on the principle of cavitation that generates shear forces to cleave the complex lignocellulosic structure, thus promoting the extraction of desired compounds such as cellulose and hemicellulose [13].

Physical pretreatment, the simplest method of delignification, is environment-friendly and seldom produces any toxic by-products [14]. However, physical methods require high energy and thus are not economically viable. Also, equipments such as sawing machines and grinders are expensive and also need specific operating expertise further increasing the operating cost [15].

4.1.2 Chemical Method

The chemical method employs the usage of chemicals to breakdown complex lignin structure in order to react with cellulose. Most common chemical methods used include alkali pretreatment, dilute acid hydrolysis, concentrated acid hydrolysis processes, ozonolysis and organosolv process [16–18]. Among these methods, alkaline pretreatment, ozonolysis and organosolv treatment methods are usually employed to eradicate lignin from lignocelluloses. These methods, though effective for lignin solubilization, are also responsible for hydrolysis of a part of the hemicellulose fraction [16]. Use of chemical methods dominates over physical or biological pretreatments because of their effectiveness in biodegradation of complex lignocellulosic materials [19]. However, high cost of the chemicals used in chemical processes is a major drawback of this method [15]. Other important disadvantages are partial cellulose degradation and production of toxic degradation by-products.

4.1.2.1 Acid Pretreatment

The mechanism of acid pretreatment is based on the distortion of hydrogen and covalent bonds holding the various components of biomass together and its effectiveness are dependent on the susceptibility of the glycosidic bonds between cellulose and hemicellulose to acid [20, 21]. Hydronium ions originating from the acid catalyst cause the hydrolysis of cellulose and hemicellulose to their respective sugar monomers [22]. The major reaction during acid hydrolysis is the breakdown of hemicellulose fraction, especially xylan which sometimes may lead to the formation of furfural and 5-hydroxymethylfurfural (HMF) because of the dehydration of xylose galactose, mannose and glucose [23]. Both organic acids such as formic acid, maleic acid, oxalic acid and inorganic acids such as nitric acid, sulphuric acid, phosphoric acid and hydrochloric acid are used for acid pretreatment [21, 23]. Acid pretreatment is performed either with concentrated acids in the range of 30–70% at a low temperature (<100 °C) or with dilute acids in the range of 0.1–10% at a high temperature (100–250 °C) depending on the type of substrate and expected results.

Mineral acids including H_2SO_4 or HCl are generally employed in concentrated acid pretreatment [17]. These harsh conditions are needed to reduce the crystallinity of cellulose and liberate glucose from the complex structure of cellulose chains. Although, the concentrated acid hydrolysis promises high sugar conversion rate, their toxic and corrosive nature demands high operational and maintenance costs. Additionally, the generation of inhibitory compounds like phenolic acids, furfurals, 5-hydroxymethylfurfural (HMF) and aldehydes on partial cellulose and hemicellulose degradation is undesirable. However, several chemical and biological detoxification methods can be employed to minimize their effect. Thus, the concentration of acid, reaction time and temperature are crucial factors and need to be optimized to avoid any sugar loss and formation of toxic compounds [20, 24].

Dilute acid pretreatment is performed at low acid concentrations and high temperature for hydrolysis of hemicellulose portion of lignocelluloses to its component sugars with minimal sugar decomposition [25]. Sulphuric acid is most commonly employed acid although other acids like HCl, HNO_3 and H_3PO_4 are also used. The dilute acid catalyzes the breakdown of long hemicellulose chains to release shorter chain oligomers which further breakdown to corresponding sugar monomers. A significant advantage of this method is the use of less severe conditions for release of hemicellulose sugars owing to the amorphous nature of hemicellulose. Another major advantage is the generation of lower degradation products. Lignin is also degraded to a high degree, resulting in increased accessibility of cellulose to the enzymes. Dilute acid pretreatment is considered to be the most suitable pretreatment used at industrial scale due to their economical and environmentally friendly manner [7, 21].

4.1.2.2 Alkali Pretreatment

Alkali pretreatment involves the application of bases to lignocellulosic biomass that results in swelling, leading to an increased internal surface area of the substrate and decreased degree of polymerization and crystallinity of cellulose fraction, distortion of lignin structure and destruction of linkages between lignin and carbohydrates [26]. Effectiveness of alkali pretreatment depends on the percentage of lignin content in the biomass as it works better with biomass of low lignin content biomass [27]. NaOH, KOH and $Ca(OH)_2$ are the most reported bases used for alkali pretreatment of lignocelluloses. Among these, NaOH has been one of the most commonly used chemicals for delignification of agro-residues and was also found to be the most effective [28]. The mechanism behind alkali pretreatment involves a saponification reaction which facilitates the cleavage of intermolecular ester bond between lignin and hemicelluloses resulting in their solubilization [29]. Furthermore, separation of acetyl groups and uronic acid from hemicellulose fragments enhances the polysaccharides accessibility to enzymatic hydrolysis [29]. Also, the lignin content on alkali pretreatment is dissolved and separated from the raw substrate as the process effluent rich in phenolic compounds [25]. The main disadvantage of this technique is that some part of the hemicellulose fragment is also degraded.

Alkaline pretreatment using calcium hydroxide has also been studied and found to be cost-effective, safer, environmental-friendly and can be easily recovered [30]. $Ca(OH)_2$ pretreatment has also been reported to enhance methane and biogas yield from lignocellulosic biomass [31]. Ammonia-based alkali pretreatment methods have also been studied extensively because of non-corrosive and non-toxic nature and easy recovery of ammonia [32, 33]. Another alkali-based pretreatment method is hydrogen peroxide treatment which is performed at high temperature (<100 °C) to promote a fast decomposition of H_2O_2. This results in the formation of reactive ions like hydroxyl radicals (HO) and superoxide anions (O^{2-}) which ultimately cause lignin degradation. This pretreatment technique is generally employed in pulp and paper industries for bleaching and delignification purposes [34].

4.1.2.3 Organosolvents Pretreatment

Organosolv is an emerging pretreatment process which uses organic solvents like acetone, ethanol, methanol, organic acids and ethylene glycol or their aqueous solutions to treat biomass. Hydrolysis of lignin and hemicellulose and also internal bonds between them by this mixture leaves relatively pure cellulose for further processing. This process of delignification and hemicellulose solubilization leads to an increase in surface area and pore size of cellulose and improves the process of enzymatic saccharification [3, 35]. The process is often performed with the supplementation of a catalyst in the form of acid (hydrochloric acid, sulphuric acid, phosphoric acid oxalic, salicylic and acetylsalicylic acid), base (lime, sodium hydroxide and ammonia) or some salt to lower the temperature requirement and enhance the rate of delignification. Major disadvantages associated with this process are expensive chemicals, energy-intensive recovery and high flammability and volatility of the organic solvents [36].

4.1.2.4 Ozone Pretreatment

Ozone pretreatment is another method used for eradication of the lignin content of lignocellulosic substrate. Since lignin degradation consumes most of the ozone during the degradation process, only low ozone amount is left for cellulose degradation, thus preserving most of it for saccharification. However, ozone may still cause some cellulose degradation by oxidation of hydroxyl groups of glucose in the presence of free radicals as mediators or by reacting directly with the glycosidic bonds of cellulose [34]. This treatment may also attack the hemicellulose component of the lignocelluloses. Ozonolysis is carried out at room temperature and normal pressure and since ozone can be easily decomposed by using a catalytic bed or increasing the temperature, the process can be designed to minimize the environmental pollution [17].

4.1.3 Physico-Chemical Pretreatment

Physico-chemical pretreatment includes steam explosion, ultrasound pretreatment, CO_2 explosion and liquid hot water (LHW) pretreatment.

4.1.3.1 Steam Explosion

Steam explosion is a method commonly used for hemicellulose hydrolysis and is typically a combination of both mechanical and chemical reactions. The biomass during steam explosion is subjected to saturated steam under high pressure (0.69–4.83 MPa) and temperature (160–260 °C) for a short period of few seconds to minutes to let the water molecules penetrate the substrate particles. The pressure is then rapidly released which makes the material undergo an explosive decompression to allow the water molecules to escape in an explosive manner [17]. This sudden release of pressure also causes the bulk lignocellulose to explode into splitted fibres. The high temperature and pressure further facilitate the cleavage of the glycosidic bonds in the carbohydrates and lignin–hemicellulose bonds [37]. Apart from glucose and xylose monomers, hemicellulose hydrolysis also liberates acetic acid which catalyzes further hydrolysis of hemicelluloses; hence, steam explosion is also referred to as autohydrolysis [30]. Several benefits of steam explosion include cost and energy effectiveness, limited chemicals use, higher sugar recovery compared to other pretreatments and no recycling costs. Its limitations are the generation of toxic inhibitory compounds and incomplete disruption of the lignin–carbohydrate matrix [20, 38].

4.1.3.2 Liquid Hot Water (LHW)

Much similar to the steam explosion process, liquid hot water (LHW) pretreatment uses water in place of steam at elevated temperatures (170–230 °C) and pressure (up to 5 MPa) over reaction time ranging from a few minutes to an hour. However, unlike steam explosion process, rapid depressurization is not required and pressure is only required to maintain its liquid form and prevent evaporation of water in order to promote disintegration and separation of the lignocellulosic matrix [35, 39]. Temperature dominates the types of sugar formation and time dominates the amount of sugar formation [39].

LHW releases the acetyl groups of hemicellulose and degrades lignin thereby enhancing the exposure of cellulose for enzymatic hydrolysis [35]. The released hemicellulose fraction remains in the liquid portion of the treated slurry and to avoid any sugar loss during this period, pH is strictly maintained between 4 and 7 during the LHW treatment process [40]. LHW pretreatment does not require any catalyst, produce minimal toxic by-product. Moreover, since the biomass particles are broken down during the course of pretreatment, substrate particle size has no impact on the efficiency of the process making it desirable for large-scale utilization. However, high demand for water and energy is its major limitation [41].

4.1.3.3 Ultrasound Pretreatment

The influence of ultrasound application on lignocelluloses has been exploited for extraction of lignin, cellulose and hemicelluloses. However, limited research has been conducted to study and explore the susceptibility of lignocellulosic biomass to enzymatic hydrolysis after ultrasonic pretreatment [42]. Ultrasound produces chemical and physical effects like delignification and surface erosion on the lignocellulosic structure which augment the pretreatment of lignocellulose. Low frequency waves together with alkaline solutions are commonly employed. The magnitude of effect of ultrasonic treatment is generally regulated by ultrasonic frequency, types of solvent and reactor type [43].

4.1.3.4 CO_2 Explosion

This method utilizes CO_2 as a supercritical fluid under high pressure to hydrolyze the carbohydrates part of the biomass. The CO_2 molecules on penetrating into the biomass shatter the highly complex lignocellulosic structure comprising holocellulose and lignin. This CO_2 when dissolved in water forms carbonic acid which in turn catalyzes the hydrolysis of hemicellulose and also the rapid release of the pressurized gas disrupts the compact lignocellulosic structure thereby improving the accessibility of cellulose to hydrolysis and leads to high glucose yield [44]. Carbon dioxide is helpful in hydrolyzing hemicellulose as well as cellulose [17]. The delignification process with CO_2 treatment can be improved by implementation of solvents such as ethanol–water or acetic acid–water [45].

Advantages such as low cost CO_2, no toxins formation, low environmental impact, non-flammability, easy recovery after extraction, lower energy requirement and greener alternative compared to ammonia makes the employment of supercritical-CO_2 an attractive approach. However, the disadvantages of the method are low yield, requirement of very high pressure and low efficiency in biomass with no moisture content [46].

4.1.3.5 Ammonia Fibre Explosion (AFEX)

The ammonia fibre explosion (AFEX) process is a physico-chemical process in which the lignocellulosic material is subjected to liquid anhydrous ammonia (1:1 ratio) in a closed vessel under conditions of high temperature (60–100 °C) and pressure for exposure time of 5–30 min, and is then rapidly depressurized [14]. The high temperature and pressure result in swelling of lignocellulose particles and the rapid depressurization distorts the complex matrix of biomass, reduces the degree of crystallinity and thereby exposes cellulose for improved accessibility to enzymes. The four significant parameters during AFEX process include temperature, pressure, water and ammonia loading [47]. However, the degree of degradation of the biomass is mostly dependent on the process temperature as it affects the speed of vaporization of ammonia during depressurization. Ammonia loading used is typically 1 kg ammonia per kg dry biomass for most of the lignocellulosic feedstocks. This pretreatment method does not remove lignin but causes the breakdown of the lignin–carbohydrate complexes [16]. The major advantage of the process is the negligible production of inhibitors. Also, the temperature used is significantly lower than steam explosion process leading to low energy input and an overall cost reduction.

4.1.4 Thermo-Chemical Processes

The thermo-chemical processes generally use much higher temperatures and pressures than other chemical treatments. Gasification is a thermo-chemical process, wherein the biomass is converted into synthesis gas, also known as syngas which is a mixture of hydrogen and carbon monoxide. Pyrolysis is another such technique in which organic material is heated in the absence of oxygen. During pyrolysis, biomass is treated at temperatures above 300 °C leading to decomposition of cellulose to gaseous products and residual char. The high temperatures usage, formation of volatile products and the cooling cost of the system are major limitations of this method. The thermo-chemical methods are used to produce a wider variety of fuels than biochemical conversion processes [48].

4.1.5 Biological Pretreatment

Biological pretreatment usually refers to the deconstruction of complex and recalcitrant lignin fraction of the lignocellulosic substrates using micro-organisms and/or ligninolytic enzymes [49]. Brown-, white- and soft-rot fungi, actinomycetes and bacteria as whole cell or enzymes which degrade lignin, the most recalcitrant polymer are used to treat lignocellulosic biomass. The lignin-degrading enzymes involved in the process include laccases (lac), lignin peroxidase (LiP), manganese peroxidase (MnP) and versatile peroxidase. White-rot, brown-rot and soft-rot fungi are most commonly used for degradation of lignin and hemicelluloses but white-rot fungi are considered to be the most efficient agents for biological pretreatment owing to high delignification rate and improved sugar yield in biological pretreated samples [50]. Unlike harsh and cost-/energy-intensive physical and chemical pretreatment methods, biological pretreatment processes are known for their mild and environment-friendly nature. Degradation of lignin by fungi, such as *Phanerochaete chrysosporium, Trametes versicolor, Trametes hirsuta* and *Bjerkandera adusta*, allows better accessibility to the cellulose and hemicellulose and is also an effective biological detoxification alternative. Major limitation associated with biological pretreatment methods is the non-selective nature of fungi which may also attack cellulose and hemicellulose. Two major families of ligninolytic enzymes considered to play key role in the process of lignin degradation include peroxidase family (lignin peroxidase, LiP and manganese peroxidase, MnP) and phenol oxidase (laccase). Other accessory enzymes that participate in lignin degradation but their specific roles remain unravelled are glyoxal oxidase, glucose oxidase, oxidoreductase and methanol oxidase [20, 49].

4.2 Inhibitory Compounds in Lignocellulosic Hydrolysate and Their Detoxification

Chemical pretreatment of lignocellulose, especially with dilute acid pretreatment method, is accompanied by the generation of numerous degradation products, a number of which are inhibitory to the growth and metabolism of microbial cells used in further stages of bioethanol production. These inhibitors belong to three main classes based on their origin: phenolic compounds, weak acids and furan derivatives [51].

Aromatic compounds formed during degradation of sugars are predominantly furan derivatives, the most prominent being furfural originating from pentoses and hydroxymethylfurfural (HMF) originating from hexoses. Furans are generated during harsh acid pretreatment conditions and are known to be the most potent inhibitors of growth and metabolism of fermentative yeast, thus affecting the process of fermentation [3, 20, 24 and 52]. Among the acid inhibitors, acetic acid is pervasive in hemicellulose hydrolysates causing acetylation of both hemicellulose and lignin. Hydrocarboxylic acids such as lactic acid and glycolic acid are produced during alkaline pretreatment. Formic acid is the product of sugar degradation, while levulinic acid is produced by 5-hydroxymethylfurfural degradation [52]. Phenolic compounds are by-products of lignin degradation during dilute acid hydrolysis. Some of these phenolics are strongly inhibitory even at relatively low concentrations while others require much higher concentrations to exert an inhibitory effect [48]. The most common phenolic compounds produced during lignin degradation are 4-hydroxybenzaldehyde, 4-hydroxybenzoic acid, coniferyl aldehyde, dihydroconiferyl alcohol, syringaldehyde, syringic acid and vanillin [52].

Various physical, chemical and biological detoxification methods have been proposed either for their transformation to inactive compounds or to reduce their concentration to levels with minimized inhibitory effect. However, the effectiveness of any of these detoxification methods is dependent on both the type of hydrolysate and on the species of micro-organisms employed. These detoxification strategies include the use of activated charcoal, extraction with organic solvents, molecular sieves, over liming, intracellular acidification, ion exchange, ion exclusion yeast strain variation and recombinant strains [53]. The employment of activated charcoal for hemicellulose hydrolysate detoxification is considered to be an effective method with low costs and high absorption capacity without affecting the sugar levels [54, 55]. A combination of high temperature and pH is also a promising method for detoxification of dilute sulphuric acid-pretreated hydrolysate of lignocellulosic biomass [24, 54]. The process has proven helpful to remove inhibitory compounds such as furfural and hydroxymethylfurfural (HMF). The efficiency of any detoxification method is dependent on the type of lignocellulosic hydrolysate, different pretreatment variables such as temperature, pH and reaction time [55]. Also, in order to promote the generation of specific, efficient and cost-effective detoxification techniques, advanced research is required for the identification of major inhibitory products and their mechanisms. A newer approach as an alternative to detoxification is the development of inhibitor-tolerant strains using genetic reconstruction and metabolic engineering.

References

1. Higuchi T (2006) Look back over the studies of lignin biochemistry. J Wood Sci 52:2–8
2. Alvira P, Tomas-Pejo E, Ballesteros M, Negro MJ (2010) Pretreatment technologies for an efficient bioethanol production process based on enzymatic hydrolysis: a review. Bioresour Technol 101:4851–4861

3. Taherzadeh MJ, Karimi K (2008) Pretreatment of lignocellulosic wastes to improve ethanol and biogas production: a review. Inter J Molecular Sci 9:1621–1651
4. Romani A, Garrote G, Alonso JL, Parajo JC (2010) Bioethanol production from hydrothermally pretreated *Eucalyptus globulus* wood. Bioresour Technol 101:8706–8712
5. Barakat A, Mayer C, Solhy A, Arancon RAD, De Vries H, Luque R, Barakat A, Mayer C, Solhy A, Arancon RAD, De Vries H (2014) Mechanical pretreatments of lignocellulosic biomass: towards facile and environmentally sound technologies for biofuels production. RSC Adv 4:48109–48127
6. Cheng JJ, Timilsina GR (2011) Status and barriers of advanced biofuel technologies: a review. Renew Energy 36:3541–3549
7. Kumar AK, Sharma S (2017) Recent updates on different methods of pretreatment of lignocellulosic feedstocks: a review. Bioresour Bioprocess 4:7. https://doi.org/10.1186/s40643-017-0137-9
8. Karunanithy C, Muthukumarappan K, Julson JL (2008) Influence of high shear bioreactor parameters on carbohydrate release from different biomasses. In: Annual international meeting. American Society of Agricultural and Biological Engineers
9. Aguilar-Reynosa A, Romani A, Rodriguez-Jasso RM, Aguilar CN, Garrote G, Ruiz HA (2017) Microwave heating processing as alternative of pretreatment in second-generation biorefinery: an overview. Energy Conver Manage 136:50–65
10. Li H, Qu Y, Yang Y, Chang S, Xu J (2016) Microwave irradiation–a green and efficient way to pretreat biomass. Bioresour Technol 199:34–41
11. Liu Y, Sun B, Zheng X, Yu L, Li J (2018) Integrated microwave and alkaline treatment for the separation between hemicelluloses and cellulose from cellulosic fibers. Bioresour Technol 247:859–863
12. Tayyab M, Noman A, Islam W, Waheed S, Arafat Y, Ali F et al (2018) Bioethanol production from lignocellulosic biomass by environmentfriendly pretreatment methods: a review. Appl Ecol Env Res 16:225–249
13. Ravindran R, Jaiswal AK (2016) A comprehensive review on pre-treatment strategy for lignocellulosic food industry waste: challenges and opportunities. Bioresour Technol 199:92–102
14. Shirkavand E, Baroutian S, Gapes DJ, Young BR (2016) Combination of fungal and physicochemical processes for lignocellulosic biomass pretreatment–a review. Renew Sust Energ Rev 54:217–234
15. Sanchez OJ, Cardona CA (2008) Trends in biotechnological production of fuel ethanol from different feedstocks: review. Bioresour Technol 99:5270–5295
16. Wyman CE, Dale BE, Elander RT, Holtzapple M, Ladisch MR, Lee YY (2005) Coordinated development of leading biomass pretreatment technologies. Bioresour Technol 96:1959–1966
17. Kumar P, Barrett DM, Delwiche MJ, Stroeve P (2009) Methods for pretreatment of lignocellulosic biomass for efficient hydrolysis and biofuel production. Indus Eng Chem Res 48:3713–3729
18. Us E, Perendeci NA (2012) Improvement of methane production from greenhouse residues: optimization of thermal and H_2SO_4 pretreatment process by experimental design. Chem Eng J 181–182:120–131
19. Zhou S, Zhang Y, Dong Y (2012) Pretreatment for biogas production by anaerobic fermentation of mixed corn stover and cow dung. Energy 46:644–648
20. Baruah J, Nath BK, Sharma R, Kumar S, Deka RC, Baruah DC, Kalita E (2018) Recent trends in the pretreatment of lignocellulosic biomass for value-added products. Front Energy Res 6:141. https://doi.org/10.3389/fenrg.2018.00141
21. Amin FR, Khalid H, Zhang H, Rahman S, Zhang R, Liu G, Chen C (2017) Pretreatment methods of lignocellulosic biomass for anaerobic digestion. AMB Expr 7:72
22. Lloyd TA, Wyman CE (2005) Combined sugar yields for dilute sulfuric acid pretreatment of corn stover followed by enzymatic hydrolysis of the remaining solids. Bioresour Technol 96:1967–1977
23. Hendriks ATWM, Zeeman G (2009) Pretreatments to enhance the digestibility of lignocellulosic biomass. Bioresouce Technol 100:10–18

24. Jonsson LJ, Martin C (2016) Pretreatment of lignocellulose: formation of inhibitory by-products and strategies for minimizing their effects. Bioresour Technol 199:103–112
25. Mussatto SI, Roberto IC (2006) Chemical characterization and liberation of pentose sugars from brewers spent grain. J Chem Technol Biotechnol 81:268–274
26. Badiei M, Asim N, Jahim JM, Sopian K (2014) Comparison of chemical pretreatment methods for cellulosic biomass. Procedia Soc Behav Sci 9:170–174. https://doi.org/10.1016/j.apcbee.2014.01.030
27. Mudhoo A (2012) Biogas production: pretreatment methods in anaerobic digestion. Scrivener Publishing, Massachusetts
28. Kim JS, Lee YY, Kim TH (2016) A review on alkaline pretreatment technology for bioconversion of lignocellulosic biomass. Bioresour Technol 199:42–48. https://doi.org/10.1016/j.biortech.2015.08.085
29. Sun FF, Zhao X, Hong J, Tang Y, Wang L, Sun H et al (2016) Industrially relevant hydrolyzability and fermentability of sugarcane bagasse improved effectively by glycerol organosolv pretreatment. Biotechnol Biofuels 9:59. https://doi.org/10.1186/s13068-016-0472-7
30. Singh J, Suhag M, Dhaka A (2015) Augmented digestion of lignocellulose by steam explosion, acid and alkaline pretreatment methods: a review. Carbohydr Polym 117:624–631
31. Shah TA, Tabassum R (2018) Enhancing biogas production from lime soaked corn cob residue. Int J Renew Energy Res 8:761–766
32. Sakuragi K, Igarashi K, Samejima M (2018) Application of ammonia pretreatment to enable enzymatic hydrolysis of hardwood biomass. Polym Degrad Stab 148:19–25
33. Oladi S, Aita GM (2017) Optimization of liquid ammonia pretreatment variables for maximum enzymatic hydrolysis yield of energy cane bagasse. Ind Crops Prod 103:122–132
34. Nigam P, Gupta N, Anthwal A (2009) Pretreatment of agro-industrial residues. In: Biotechnology for agro-industrial residues utilisation. Springer, Netherlands, 13–33
35. Zhang K, Pei Z, Wang D (2016) Organic solvent pretreatment of lignocellulosic biomass for biofuels and biochemicals: a review. Bioresour Technol 199:21–33
36. Borand MN, Karaosmanoglu F (2018) Effects of organosolv pretreatment conditions for lignocellulosic biomass in biorefinery applications: a review. J Renew Sustain Ener 10:033–104
37. Chen HZ, Liu ZH (2015) Steam explosion and its combinatorial pretreatment refining technology of plant biomass to bio-based products. Biotechnol J 10:866–885
38. Pielhop T, Amgarten J, Rohr PR, Studer MH (2016) Steam explosion pretreatment of softwood: the effect of the explosive decompression on enzymatic digestibility. Biotechnol Biofuels 9:152
39. Yu G, Yano S, Inoue H, Inoue S, Endo T, Sawayama S (2010) Pretreatment of rice straw by a hot-compressed water process for enzymatic hydrolysis. Appl Biochem Biotechnol 160(2):539–551
40. Li HQ, Jiang W, Jia JX, Xu J (2014) pH pre-corrected liquid hot water pretreatment on corn stover with high hemicellulose recovery and low inhibitors formation. Bioresour Technol 153:292–299
41. Bhutto AW, Qureshi K, Harijan K, Abro R, Abbas T, Bazmi AA et al (2017) Insight into progress in pre-treatment of lignocellulosic biomass. Energy 122:724–745
42. Sun RC, Tomkinson RC (2002) Characterization of hemicelluloses obtained by classical and ultrasonically assisted extractions from wheat straw. Carbohyd Polym 50:263–271
43. Bussemaker MJ, Zhang D (2013) Effect of ultrasound on lignocellulosic biomass as a pretreatment for biorefinery and biofuel applications. Indus Energy Chem Res 52(10):3563–3580
44. Capolupo L, Faraco V (2016) Green methods of lignocellulose pretreatment for biorefinery development. Appl Microbiol Biotechnol 100:9451–9467
45. Pasquini D, Pimenta MTB, Ferreira LH, Curvelo AADS (2005) Extraction of lignin from sugarcane bagasse and *Pinus taeda* wood chips using ethanol-water mixtures and carbon dioxide at high pressures. J Supercrit Fluid 36:31–39
46. Kim KH, Hong J (2001) Supercritical CO2 pretreatment of lignocellulose enhances enzymatic cellulose hydrolysis. Bioresour Technol 77:139–144
47. El-Naggar NE, Deraz S, Khalil A (2014) Bioethanol production from lignocellulosic feedstocks based on enzymatic hydrolysis: current status and recent developments. Biotechnology 13:1–21

48. Larson ED (2008) Biofuel production technologies: status, prospects and implications for trade and development. In: United Nations conference on trade and development (UNCTAD)
49. Tanjore D, Richard TL (2015) A systems view of lignocellulose hydrolysis. In: Ravindra P (ed) Advances in bioprocess technology. Springer International Publishing, Cham, pp 387–419
50. Yesilada O, Birhanli E, Geckil H (2018) Bioremediation and decolorization of textile dyes by white rot fungi and laccase enzymes. In: mycoremediation and environmental sustainability, fungal biology. Springer, Cham, 121–153
51. Liu ZL (2006) Genomic adaptation of ethanologenic yeast to biomass conversion inhibitors. Appl Microbiol Biotechnol 73:27–36
52. Klinke HB, Thomsen AB, Ahring BK (2004) Inhibition of ethanol-producing yeast and bacteria by degradation products produced during pretreatment of biomass. Appl Microbiol Biotechnol 66:10–26
53. Rao RS, Jyothi CP, Prakasham RS, Sarma PN, Rao LV (2006) Xylitol production from corn fibre and sugarcane bagasse hydrolysates by *Candida tropicalis*. Bioresour Technol 97:1974–1978
54. Chandel AK, Kapoor RK, Singh AK, Kuhad RC (2007) Detoxification of sugarcane bagasse hydrolysate improves ethanol production by *Candida shehatae* NCIM 3501. Bioresour Technol 98:1947–1950
55. Carvalho GB, Mussatto SI, Candido EJ, e Silva J JA (2006) Comparison of different procedures for the detoxification of eucalyptus hemicellulosic hydrolysate for use in fermentative processes. J Chem Technol Biotechnol 81:152–157

Chapter 5
Biological Pretreatment: Need of the Future

Abstract Pretreatment is one of the most significant steps for the production of bioethanol from lignocellulosic feedstocks. Among the various pretreatment strategies known, biological pretreatment seems to be the most promising as it is mild and eco-friendly process and does not generate inhibitor unlike physical and chemical pretreatment methods which are harsh and cost and energy intensive. However, there are a few limitations associated with this strategy to be used on a pilot scale. These are the long residence time for effective delignification and the non-selective nature of fungi which may also attack cellulose and hemicellulose. Nevertheless, the complete potential of biological pretreatment has not been fully exploited yet. Hence, there is a need for extensive research in the field of biological pretreatment for the development of an economically feasible process. This chapter presents an overview of various aspects of biological pretreatment including organisms responsible and mechanism, methods of delignification and various parameters affecting biological pretreatment process.

Keywords Basidiomycetes · Biological · Delignification · Laccase · Lignin · Pretreatment

5.1 Lignocellulose-Degrading Organisms

A wide range of micro-organisms including fungi, bacteria and actinomycetes are capable of lignin degradation. The complete and efficient degradation of lignin is generally achieved as a result of the interaction of certain fungi and bacteria, in which fungi play a major role. The bioconversion of lignocelluloses is predominantly carried out by fungi, and white-rot fungi appear to be the most efficient lignin degraders. Two major types of fungal enzymes known to break down lignocellulosic material include an oxidative and extracellular ligninolytic system for lignin degradation and the hydrolytic system that produces various hydrolases responsible for degradation of polysaccharide fraction. The ability of these organisms to produce specific enzymes for degradation of different lignocellulosic substrates is due to the diverse habitat

A. Sharma and N. K. Aggarwal, *Water Hyacinth: A Potential Lignocellulosic Biomass for Bioethanol*, https://doi.org/10.1007/978-3-030-35632-3_5

they are found. Table 5.1 describes various anatomical and chemical characteristics associated with lignin degradation by white-rot, soft-rot and brown-rot fungi and bacteria.

Table 5.1 Anatomical and chemical characteristics of lignin-degrading and wood-decaying organisms

Features	White rot fungi		Brown rot fungi	Soft rot fungi	Bacteria
Decay appearance and consistency	Bleached appearance, lighter in colour than sound wood, moist, soft, spongy, strength loss after advanced decay		Brown, dry, crumbly, powdery, brittle consistency, breaks up like cubes, drastic loss of strength at initial stage of decay. Very uniform ontogeny of wood decay	Soft consistency in wet environments. Brown and crumbly in dry environments. Generally uniform ontogeny of wood decay	Soft consistency
	Simultaneous rot	Selective delignification			
Host (wood type)	Hardwood, rarely softwood	Hardwod and softwood	Softwoods; seldom hardwoods. Forest ecosystems and wood in service	Generally hardwoods (softwoods very slightly degraded). Forest ecosystems, waterlogged woods, historic archaeological wood	Sapwood, water saturated wood, wood at late stage of decomposition, plant litter
Cell wall constituents degraded	Cellulose, lignin and hemicellulose. Brittle fracture. Fibrous feature	Initial attack selective for hemicelluloses and lignin, later cellulose also. Fibrous feature	Cellulose, hemicelluloses. Lignin slightly modified	Cellulose and hemicelluloses, lignin slightly altered	Limited lignin degradation, low-molecular weight part of lignin and degradation products
Anatomical features	Cell wall attacked progressively from lumen. Erosion furrows associated with hyphae	Lignin degradation in middle lamella and secondary wall. Middle lamella dissolved by diffusion mechanism (not in contact with hyphae), radial cavities in cell wall	Degradation at a great distance from hyphae (diffusion mechanism). Entire cell wall attacked rapidly with cracks and clefts	Cell wall attack in the proximity of hyphae starts from cell lumen. Longitudinal biconical cylindrical cavities in secondary wall (type 1). Secondary wall erosions from cell lumen (type 2)	Erosion bacteria grow towards the middle lamella of the wood cells and cause erosion of the fibre wall, while tunnelling bacteria grow within the cell wall

(continued)

Table 5.1 (continued)

Features	White rot fungi		Brown rot fungi	Soft rot fungi	Bacteria
Examples of fungal agents	Basidiomycetes (e.g. *T. versicolor, Irpex lacteus, P. chrysosporium* and *Heterobasidium annosum*) and some Ascomycetes (e.g. *Xylaria hypoxylon*)	Basidiomycetes (e.g. *Ganoderma australe, Phlebia tremellosa, C. subvermispora, Pleurotus spp.* and *Phellinus pini*)	Basidiomycetes exclusively (e.g. *C. puteana, Gloeophyllum trabeum, Laetiporus sulphureus, Piptoporus betulinus, Postia placenta* and *Serpula lacrimans*)	Ascomycetes (*Chaetomium globosum, Ustulina deusta*) and Deuteromycetes (*Alternaria alternata, Thielavia terrestris, Paecilomyces* sp.)	*Streptomyces, Nocardia, Pseudomonas sp., Bacillus megatarium* and *Serratia marcescens*

Source [1–5]

5.2 Lignin-Degrading Fungi

White-rot, soft-rot and brown-rot fungi are the main groups of fungi involved in lignin degradation. However, brown-rot and soft-rot fungi have a tendency to degrade cellulose along with lignin degradation [6]. While some white-rot fungi can simultaneously degrade lignin and polysaccharides resulting in the loss of carbohydrates, others can selectively degrade lignin. These white-rot fungi are the most common organisms to be used as an agent of delignification as they can effectively mineralize lignin without causing any severe damage to the cellulose content during the process. Some common examples of such lignin-selective fungi as given in Table 5.2 are *P. chrysosporium, C. Subvermispora, Pleurotus ostreatus, Phlebia radiata, Phlebia*

Table 5.2 Lignin degradation by fungi

Name of organisms	Substrate	Results
Phanerochaete chrysosporium	Polymeric dyes	Dye decolourization due to ligninolytic activity [20]
Merulius tremellosus	Aspen wood	52% lignin loss in 8 weeks [30]
Fusarium proliferatum	Polymeric kraft lignin, polymeric organosolv lignin, milled wood lignin	Residual Klason lignin content was 15% in PKL, 10% in POL, 13% in MWL, respectively, in 30 days [31]
Phlebia sp.	Sugarcane bagasse	>50% delignification [32]
Dichomitus squalens *Ceriporiopsis Subvermispora* *Pleurotus ostreatus*	*Fagus crenata* wood chips	21% lignin loss 8% lignin loss 10.8% lignin loss [33]

(continued)

Table 5.2 (continued)

Name of organisms	Substrate	Results
Pleurotus sp., *Lentinus edodes*	Milled tree leaves, banana peel, apple peel, mandarin peel	Laccase activity of 119, 195, 794 and 4103 U/l, respectively, in 3–22 days [34]
Pleurotus ostreatus and P. sajor-caju	Water hyacinth	26.2 and 30.4% delignification in 7 weeks [35]
Trametes versicolor G20	Bamboo culms	12% loss of lignin content in 120 days [36]
Phanerochaete chrysosporium	Cotton stalk	75% loss in lignin content [37]
Aspergillus terreus, Zymomonas mobilis, Aspergillus awamori	Sugarcane trash	Lignin degradation [38]
Phanerochaete chrysosporium	Cotton stalks	27.7% lignin loss and 0.027 g/g ethanol yield [23]
Irpex lacteus	Wheat straw	>50% delignification [39]
Ceriporiopsis subvermispora	Corn stover	39.2% delignification in 42 days [40]
Irpex lacteus	Corn stalks	Lignin loss was 11.84 and 27.81% in untreated and NaOH-pretreated corn stalks, respectively, in 15 days [17]
Phanerochaete chrysosporium	Water hyacinth	67.21% loss of lignin content in 8 weeks [41]
Phanerochaete chrysosporium, Ganoderma applanatum, Lenzites adusta, Dadela flavida	Bagasse, rice straw and wheat straw	Lignin loss 18–36% for bagasse, 21–34% for wheat straw and 29–39% for rice straw after 28 days of incubation [42]
Alternaria sp. *NITDS1*	Water hyacinth	50% lignin degradation [43]
Phanerochaete chrysosporium	Water hyacinth	42.44% lignin degradation in 16 days [44]
Ceriporiopsis subvermispora	Sugarcane bagasse	47% sugar was recovered as sugar-rich syrup [45]
Pleurotus florida	Paddy straw	Maximum saccharification efficiency up to 75.3% [46]

tremellosus, *Phlebia subserialis*, *Phellinus pini* and *Dichomitus squalens* [2, 7–9]. *Phanerochaete chrysosporium* has been potentially exploited as the model organism for lignin degradation studies by white-rot fungi [10–12]. Lignin degradation is mostly brought about by fungi belonging to Basidiomycetes. More than 600 species of Basidiomycetes have ligninolytic potential and capable of secreting extracellular laccase, lignin peroxidase and manganese peroxidase enzymes. Some important lignin-degrading fungi belonging to basidiomycete group are *Agaricus, Arthrobotrys, Cephalosporium, Clavaria, Clitocybecollybia, Collybia, Flammula, Fusarium, Humicola, Hypholoma, Lepiota, Mycena, Poria, Pholiota, Pleurotus* and *Polyporus* [13].

There are more than 10,000 different species of white-rot fungi in nature. More than 400 species and strains of white-rot fungi were screened for selective removal of lignin using woodblock methods, and a literature of about 72 lignin-degrading fungi was published which pointed to the dearth of knowledge about the lignin degradation by fungi [14]. The pattern of lignin removal from two softwoods (*Pinus densiflora and Cryptomeria Japonica*) and one hardwood (*Fagus crenata*) by a soft-rot fungus (*Chaetomium globosum* Kunze) and a white-rot fungus (*Coriolus versicolor* QUEL) was studied, and the rate of lignin removal was observed to be always slower in *C. globosum* than in *C. versicolor* [15]. In another study, blocks of wood were treated with *Coriolus versicolor, Dichomitus squalens, Phellinus pini, Phlebia tremellosus, Poria medulla-panis, Scytinostroma galactinum* and five different strains of *Phanerochaete chrysosporium* for 12 weeks and a 73% and 89% of lignin loss was observed with strain *Phanerochaete chrysosporium* BKM-F-1767 and *Scytinostroma galactinum* [4].

White-rot fungi degrade lignin by means of a family of extracellular oxidative enzymes collectively termed "lignases" [16]. It has been suggested that the initial attack on lignin must be carried out by extracellular and non-specific enzymes because of the complex structure and size of the lignin molecule [10, 16]. The three best studied extracellular ligninolytic enzymes in white-rot fungi are laccase, lignin peroxidases (LiPs) and manganese peroxidases (MnPs). Different white-rot fungi are known to produce different combinations of these enzymes: fungi producing LiP and MnP, fungi producing MnP and laccase, fungi producing LiP and laccase and fungi which produce neither LiP nor MnP, but laccase and aryl alcohol oxidase (AAO) or some other accessory enzyme [16–18]. *P. chrysosporium, P. radiata* and *T. versicolor* belong to the peroxidase group and produce both MnP and LiP enzymes. Usually, some of the members of peroxidase group also produce some amount of laccase with the general exception of *P. chrysosporium*, although laccase production by *P. chrysosporium* has been reported under specific conditions [19]. Most of the fungi belonging to the peroxidase and the MnP-lac groups are found to be more efficient and selective lignin degraders, while the fungi belonging to the LiP-lac group have comparatively lower lignin degradation potential [16]. Fungal peroxidases including both lignin and manganese peroxidase used in the biodegradation of lignin were discovered in *Phanerochaete chrysosporium* for the first time [20, 21]. Based on this discovery, biological pretreatment of wood and paddy straw using *Phanerochaete chrysosporium* was conducted but the degree of delignification was found to be insufficient to expose a significant fraction of cellulose for enzymatic hydrolysis [21]. Biological treatment of wheat straw using fungal isolate, RCK-1 having high lignin-degrading potential, resulted in a reduced concentration of acid loading required for chemical hydrolysis, an increase in fermentable sugar yield and a reduced production of fermentation inhibitors [22]. The potential of *Phanerochaete chrysosporium* to degrade lignin and facilitate fuel ethanol production from cotton stalks was studied. The biological pretreatment of cotton stalks by *Phanerochaete chrysosporium* resulted in significant reduction in lignin and hemicellulose content compared to the untreated stalks [23].

White-rot fungi are the organisms responsible for major part of lignin decomposition in nature but their growth substrate also hemicelluloses and cellulose content and not just lignin [1, 4, 10]. White-rot fungi can cause both selective delignification and non-selective delignification of wood. Among the best studied white-rot fungi, *Phanerochaete chrysosporium* and *Phlebia radiata* are selective lignin degraders while *Trametes versicolor* is non-selective in nature [2, 4, 16]. The mineralization of ^{14}C-lignin under optimal environmental conditions by the most selective white-rot fungi can reach up to 50% or even 70% [16]. Chemical modification of lignin is attained by demethylation of its phenolic and non-phenolic units, limited aromatic hydroxylation and ring cleavage [2, 10]. Brown-rot fungi are able to efficiently degrade methoxyl groups of the lignin with limited mineralization of other parts [1, 10]. Brown-rot fungi mainly colonize softwoods and consist of only 6% of all wood-rotting Basidiomycetes. Soft-rot fungi including fungal species from both Ascomycetes and Deuteromycetes can degrade both hardwood and softwood, but hardwoods are degraded to a greater extent. Although soft-rot fungi are capable of degrading all wood components, the rate of decomposition is much lower compared to white-rot and brown-rot fungi [2]. Examples of soft-rot fungi involved in lignin degradation are *Chaetomium piluliferum*, *Penicillium chrysogenum, Fusarium solani* and *Fusarium oxysporum* [24, 25]. Soft-rot fungi have also been known to degrade lignin-related compounds, like vanillic acid and phenols, but little is known about their enzymatic system or lignin degradation capacity as litter decomposing organisms [10, 24].

Although Basidiomycetes are the primary wood decay fungi, other groups of fungi are also involved in the lignin decomposition processes. Some higher Ascomycetes, particularly *Daldinia concentrica*, are known to degrade aspen wood with intensity comparable to *Trametes versicolor*, a basidiomycete. Yet another ascomycete, *Chrysonilia sitophila*, was found to be capable of degrading rice hull and *Pinus radiata* bark products and produced both ligninolytic and cellulolytic enzymes [26]. Various *Penicillium* strains were found to degrade compounds structurally similar to lignin and thus were described as potential lignin degraders [25]. The family Xylariaceae had been reported to produce laccase and low amounts of peroxidases along with some polysaccharide degrading hydrolases [27]. Ascomycetous fungi like *Coniothyrium minitans, Coniothyrium sporulosum* and *Coniothyrium cerealis* were found to colonize and degrade wood [28]. *Paraconiothyrium variabile* isolated from decayed wood of *Salix matsudana Koidzin* was found to utilize several lignin model compounds, decolorize aromatic dyes and degrade woodblocks causing a 22.99% loss in lignin content demonstrating higher degradation ability for lignin [29].

5.3 Lignin-Degrading Bacteria

Bacteria can only degrade the low molecular weight part of lignin and the degradation products of lignin [2, 4]. Thus, the role of bacteria may be significant in degrading the low molecular weight intermediate degradation products produced by fungal pretreatment. There are many genera of eubacteria and actinomycetes which can degrade and assimilate extracted lignin [1, 47]. However, wood and other lignocelluloses are only degraded by bacteria under specific environmental conditions like wood saturated with water, wood with a high extractive content or in anaerobic conditions but at a very slow rate [2, 4]. An erosion bacterium causes erosion of the fibre wall, while the tunnelling bacteria grow inside the cell wall [2]. *Pseudomonas, Alcaligenes, Arthrobacter, Nocardia* and *Streptomyces* are some of the genera capable of degrading single-ring aromatic compounds. Some examples of lignin-degrading bacteria and their reactions on different substrates are discussed in Table 5.3. Although aerobic bacteria are known to be the primary lignin degraders in most environments, studies suggest that anaerobic rumen bacteria can degrade plant fibre cell walls. It has also been reported that the intermediate metabolic product of aerobic degradation, oligolignols, can further be converted to CO_2 and CH_4 by anaerobic micro-organisms [48].

Among these eubacteria, *Pseudomonas* sp. is the most efficient lignin degrader. Several bacteria isolated from compost soil, viz. *Azotobacter*, *Bacillus megaterium* and *Serratia marcescens*, were found to be potential lignin degraders [49]. Another study reported laccase activity from *S. marcescens* along with lignin mineralization and solubilization [50]. *Bacillus* sp. isolated from Egyptian soil was tested for biodegradation of lignin alkali, and its molecular and physiological characteristics were studied in order to maximize the bacterial lignin degradation [51].

Table 5.3 Lignin degradation by bacteria

Bacteria species	Substrate and reaction
Streptomyces viridosporus T7A	Depolymerize lignin [53]
Azotobacter sp. HM121	Lignin mineralization and solubilization [49]
Paenibacillus sp.	Degrade kraft lignin [54]
Cellulomonas cartea, Cellulomonas uda, Bacillus macerans, Zymomonas mobilis	5–8% loss in lignin content of sugarcane trash [38]
Bacillus sp.	Degrade 81.6% of kraft lignin [55]
Pandoraea norimbergensis, Pseudomonas sp., *Bacillus* sp.	Degradation of indicator dyes (azure B, methylene blue, congo red, toluidine blue, RBBR) [56]
Streptomyces sp. strains F-6	Degrade 37.6% of alkali lignin [57]
Pandoraea sp. B-6	Degrade 46.5% of kraft lignin [58]
Citrobacter freundii	Degrade 49% of kraft lignin [59]
Bacillus sp. strains CS-1	Degrade 61% of alkali lignin [60]

Filamentous bacteria belonging to the genus Streptomyces are well-known lignin degraders. Although they can solubilize lignin and cause structural modifications, their lignin mineralization ability is quite limited [1, 2, 47]. Actinomycetes degrade lignin during primary metabolism and at high nitrogen levels compared to white-rot fungi, which degrade lignin during their secondary metabolism and in nitrogen-limited cultures. *Streptomyces badius* was found to mineralize 3.2% of ^{14}C-MWL in 30 days, and the mineralization was further enhanced to 11% on supplementing the media with cellulose and yeast extract [2]. It has been evidenced that grass lignin is attacked more efficiently by actinomycetes [1, 52].

5.4 Methods of Delignification

During biological delignification process, delignification can be achieved using four different methods depending on the application of ligninolytic enzymes. These four methods are as follows:

(1) Fungal delignification, (2) enzymatic delignification, (3) laccase–mediator system (LMS) and (4) integrated fungal fermentation (IFF).

5.4.1 Fungal Delignification

Fungal delignification utilizes the whole cell micro-organism for biomass delignification. The micro-organism can be cultivated with the biomass to be degraded either in submerged culture [61] or in solid-state fermentation [7, 40]. The organism secretes one or more extracellular ligninolytic enzymes responsible for the degradation of lignin in the culture medium. Twelve white-rot fungi, namely *Phanerochaete chrysosporium, Ischnoderma resinosum, Bondarzewia berkeleyi, Coriolus versicolor, Agrocybe aegerita, Dichomitus squalens, Flammulina velutipes, Ganoderma applanatum, Pycnoporus sanguineus, Sporotrichum pulverulentum, Coprinus fimetarius* and *Pleurotus flabellatu*, were grown in solid-state culture of lemon grass and citronella bagasse for 5–6 weeks at 20 °C, and a decrease in lignin content of up to 36% of the original value for lemon grass and 28% for citronella was observed [62]. *Eucalyptus globulus* was treated with white-rot fungi, *Coriolus versicolor*, and the changes in physical and chemical properties were determined. Loss in weight, Klason lignin content, holocellulose and cellulose was reported as 16.66, 5.35, 11.06 and 8.01%, respectively, after 30 days [63]. The degradation of synthetic lignin (^{14}C-ring-labelled DHP) in a straw-to-soil ratio of 1:5 by *Abortiporus biennis, Phlebia radiata, Phanerochaete chrysosporium, Phanerochaete sordida, Pleurotus ostreatus, Bjerkandera adusta, Dichomitus squalens, Trametes hirsuta* and *Trametes versicolor* was investigated, and ^{14}C-DHP mineralization of 4 and 23% was observed with *P. sordida* and *T. versicolor*, respectively [64]. In a study, popular wood was inoculated with cultures of *Pycnoporus sanguineus* and *Ganoderma lucidum* and found that the sample inoculated with *Pycnoporus sanguineus* showed selective lignin degradation whereas sample inoculated with *Ganoderma lucidum* exhibited simultaneous decay and selective lignin degradation [65]. A study revealed that *P. chrysosporium* when incubated on wheat straw for six weeks individually caused 28.95% lignin loss whereas up to 36% lignin loss when combined with *D. flavida* [66]. Samples of *Pinus radiata* logs were pretreated with the white-rot fungi *Ceriporiopsis* sp., *Phlebia* sp. and *Pleurotus* sp., under field condition for 90 days, and the lignin content was determined using gravimetric techniques (Klason lignin), FTIR spectroscopy and gas chromatography. The best ability to degrade lignin was observed in *Phlebia* sp. followed by *Ceriporiopsis* sp. and *Pleurotus* sp. [67].

Penicillium simplicissimum H5 when inoculated on rice straw supplemented with 1% glucose and 0.1% ammonium nitrate under solid-state fermentation with initial moisture content of 300% w/v and incubated at 30 °C was found to efficiently degrade lignin [68]. An excellent work on pretreatment of spruce wood by *Trametes versicolor* and *Bjerkandera strain* BOS55 with and without glucose supplementation indicated that *Trametes versicolor* had a higher potential to degrade spruce wood [69]. The fungal delignification of lignocellulosic feedstocks, *Prosopis juliflora* and *Lantana camara*, was carried out using cultures of *Pycnoporus cinnabarinus* ATCC 2004378 under solid-state fermentation (SSF). The results indicated more amount of lignin loss in *P. juliflora* (13.13%) than in *L. camara* (8.87%) [70]. Twenty white-rot fungi were used for screening of ligninolytic potential using bagasse, rice straw and wheat straw. Among these, five fungal strains, viz. *Phanerochaete chrysosporium, Ganoderma applanatum, Lenzites adusta, Lenzites acuta* and *Dadela flavida*, were found to be efficient lignin degraders with *Phanerochaete chrysosporium* showing maximum delignification of up to 40% [42]. Fungi showing ligninolytic potential were isolated from agro-residues of evergreen forest Kodagu D, and about 58% of the total fungal isolates demonstrated ligninolytic enzyme activity. These isolates when further tested for their capability to degrade areca and coffee husk, four isolates namely *Chaetomium* sp., *Penicillium* sp., *Aspergillus* sp. and *Trichoderma* sp., were found to degrade areca and coffee husk efficiently [71].

Lignin degradation of water hyacinth biomass by three different fungal strains *Pleurotus sajor-caju*, *P. ostreatus* and *Calocybe indica* was investigated, and the highest delignification was attained by *Pleurotus sajor-caju* (14.2%) followed by *P. ostreatus* (11.3%) and *Calocybe indica* (9.3%) [72]. Two *Pleurotus* species when used to ferment water hyacinth (*Eichhornia crassipes*) biomass were found to delignify 26.2 and 30.4% of the lignin content, respectively, and degraded most of the hemicellulose fraction after 7 weeks of incubation under solid-state fermentation [35]. Biological delignification of water hyacinth biomass with *Phanerochaete chrysosporium* using solid-state fermentation was carried out, and a loss of 34.67% of total weight and 67.21% of lignin was obtained after 8 weeks of incubation [41].

Alterneria alternata NITDS1 was studied for production of extracellular ligninolytic enzymes using water hyacinth biomass under solid-state fermentation and was found to be a potential ligninolytic enzyme producer and an efficient lignin degrader [43]. Another study on biological pretreatment of water hyacinth using *Phanerochaete chrysosporium* reported lignin degradation and fermentable sugar yield as 24.39 and 36.37% after 16 days of incubation. On supplementation of medium with cofactor Mn^{2+} at a concentration of 0.5%, delignification and the fermentable sugar yield increased to 42.44% and 67.66%, respectively [44]. These reports indicate that a wide variety of fungal strains are capable of achieving high delignification however, the rate of lignin degradation is not always proportional to high glucose yield and the loss of cellulose and hemicellulose content during fungal pretreatment is a major drawback. Lignin loss (>50%) in sugarcane bagasse by *Phlebia* sp. was accompanied by a 10% holocellulose loss [32]. In another study, *Eucalyptus grandis* wood chips were subjected to solid-state fermentation with cultures of *Ceriporiopsis subvermispora* for 15–90 days [73]. The maximum lignin loss observed was 27% after 90 days of treatment, but the cellulose content decreased from 44 to 37%, indicating that cellulose depolymerization also took place. In a study on enzymatic hydrolysis, the fungal species with better glucose concentrations were found to be different from the ones with the highest delignification rate [7]. The main challenge associated with fungal pretreatment is to improve the selectivity of fungal strains for preferential lignin degradation by application of cellulase-deficient or non-cellulose utilizing fungi, thus preserving more cellulose for enzymatic hydrolysis. Fungi with high lignin degradation specificity includes *Phanerochaete chrysosporium* [37], *Echinodontium taxodii* [17], *Irpex lacteus* [39] and *Pycnoporus sanguineus* [61], while significant amount of simultaneous carbohydrate decomposition has been reported from *Pleurotus sajor-caju*, *Ceriporiopsis subvermispora* [40], *Phlebia brevispora*, *Phlebia floridensis* and *Phlebia radiata* [74]. The total sugar yield on combination of hydrogen peroxide treatment and fungal treatment of corn stalk with *Irpex lacteus* CD2 was higher compared to the sole fungal pretreatment [75]. Another limitation associated with fungal delignification is the long residence time compared to other chemical pretreatment techniques. The residence time to obtain high delignification ranged from 13 days to 40 or 50 days [61]. The effectiveness of microbial delignification has been greatly improved by application of alkali pretreatment before subjecting the biomass to fungal treatment. This resulted in the reduction of the process duration and an increase in the glucose and ethanol yield [76]. Besides, the effect of co-culturing on rate of biodelignification has also been observed. Co-culturing of white-rot fungi *Ceriporiopsis subvermispora, Physisporinus rivulosus, Phanerochaete chrysosporium* and *Pleurotus ostreatus*, for biopulping processes, was found to enhance lignin degradation [77]. In similar work on co-culturing of white-rot fungi, an increase in the production of lignin-degrading enzymes and the degradation of the polymeric dye Poly R-478 was reported [78].

5.4.2 Enzymatic Delignification

Enzymatic delignification is the application purified or semi-purified ligninolytic enzymes or crude enzyme extract to achieve lignin degradation. The process of enzymatic delignification can employ either commercial or crude ligninolytic enzyme preparation. Laccase is the most commonly used enzyme followed by MnP and LiP. Combinations of two or three ligninolytic enzymes have also been exploited, and the synergetic relationship between the ligninolytic enzymes was found to improve delignification rate. Crude extract containing xylanase and laccase enzymes obtained from *Penicillium oxalicum* and *Pleurotus ostreatus* resulted in 21% lignin degradation in wood pulp after 3 h [79]. In a study, laccase from *Pleurotus* sp. at 400 IU/mL was used to treat *Bambusa bambos* and a significant lignin loss of 84% was achieved in 8 h [80]. Another study reported 50% delignification of wood biomass after 24 h when laccase in ionic liquid aqueous media was used for treatment [81]. Application of ligninolytic enzymes has also been reported to generate a reduced cellulose conversion in wheat straw, and thus, their deactivation prior to the saccharification process is required [82]. As stated for fungal delignification, use of alkali pretreatment method has also been employed prior to enzymatic delignification in order to facilitate the process of lignin degradation [82]. Enzymatic processes achieve the similar degree of delignification when compared to fungal pretreatment in rather less amount of time (between 24 and 96 h). Although enzymatic delignification reduces the residence time to a great extent, the current methodology is still incapable of competing with the conventional fungal pretreatment technologies in terms of cost-effectiveness.

5.4.3 Laccase–Mediator System (LMS)

LMS utilizes fungal laccases along with redox mediators to achieve efficient lignin degradation. Redox mediators are compounds that act as electron carriers between the enzyme and substrate. Laccase oxidizes the mediators increasing their redox potential, and this increment in the redox potential of mediators facilitates the degradation of lignin and other aromatic compounds that the enzyme cannot hydrolyse by itself. Electrons lost during the process are recuperated by the reaction between oxidized mediators and the substrate [83]. The most commonly used mediator compounds are 1-hydroxybenzotriazole (HBT), 2, 2′-azino-bis (3-ethylbenzo-thiazoline-6-sulphonic acid) (ABTS), syringaldehyde and vanillin [83, 84]. The principal applications of LMS are bleaching in the pulp and paper industry and delignification of feedstocks in biofuel industry.

Corn stover was treated with LMS pretreatment using HBT as a mediator, and a correlation between enzyme concentration and enzymatic digestibility of corn stover was discovered [84]. LMS pretreatment of wheat straw was carried out after the steam explosion pretreatment prior to and simultaneously with the enzymatic hydrolysis step, and the results were found to be similar in both processes. This suggested that the delignification using LMS method can be carried out simultaneously with the process of saccharification [85]. A commercial LMS procured from Genencor (Rochester, NY, USA) was used to perform delignification of cotton gin trash after alkali ultrasonication and hot water pretreatment. Application of alkali ultrasonication method enhanced the ethanol and cellulose yield at a significant level compared to just LMS treatment [86]. The major drawback associated with the application of LMS is the lack of specific redox mediators for delignification.

5.4.4 Integrated Fungal Fermentation (IFF)

Integrated fungal fermentation (IFF) is a consolidated process in which a single fungal strain or a group of fungal strains brings about the transformation of the lignocellulosic biomass into ethanol in a single stage without participating in different stages or treatments. *Phlebia* sp. MG-60 was found to be the only one from a group of 12 fungi to selectively degrade lignin and produce ethanol from this delignified biomass under semi-aerobic conditions of submerged fermentation [87]. *Phlebia* sp. MG-60 has since been evaluated for ethanol production from various feedstocks such as sugarcane bagasse, hardwood, kraft pulp, waste newspaper and agro-industrial waste [76, 87]. The delignification process and ethanol yield were further enhanced on supplementing the culture media with organic compounds and minerals, and an ethanol yield between 30 and 70% was achieved in 6–29 days depending on the type of biomass utilized [76]. The process time can however be reduced to as less as 240 h using additional pretreatment methods like alkali pretreatment [76]. Additionally, fungal co-culturing is another alternative method to generate an integrated process involving pretreatment, enzymatic saccharification and fermentation steps for ethanol production. Co-culturing using *Coprinus comatus* and *Trametes reesei* was employed for ethanol production from corn stover and enhanced lignin degradation, and cellulose conversion rate was observed compared to monoculture [88].

The process of delignification using ligninolytic enzymes can be reformed by upgrading the enzymatic catalysis process with advanced methods like protein engineering. Three different approaches for modification of ligninolytic enzymes in protein engineering are rational and semi-rational approach, and directed evolution method [89]. Rational approach implies the molecular reconstruction of the enzyme sequence by specific site direct mutation using initial structural information for the specific sequence modification. This strategy has been applied to upgrade the laccase enzymes' potential to degrade non-phenolic and high molecular weight phenolic compounds. Semi-rational approach utilizes saturation mutagenesis strategy to alter the hotspot residues in the enzymes. This technique has the ability to replace all the selected amino acids by generating additional 19 amino acids [89]. This resulted in the production of enzymes with 3-8-fold greater catalytic efficiency. Directed evolution approach employs random mutation, genetic recombination and finally selection to produce enzymes with enhanced catalytic potential and solvent tolerance [89]. The major advantages observed in fungal pretreatment method are high delignification and detoxification and the possibility of developing a consolidated process with no negative environment impact. However, the major limitations of this process are its long residence time and low efficiency. Enzymatic delignification on the other hand is highly efficient process but has high costs and low commercial availability of enzymes and specific redox mediator requirements.

5.5 Factors Affecting Production of Ligninolytic Enzymes and Lignin Degradation

A number of parameters affecting ligninolytic enzyme production and biological pretreatment include fungal strain, substrate particle size, concentration and source of carbon and nitrogen, initial moisture content, oxygen concentration, pH, temperature and incubation time [10, 41, 44].

5.5.1 Fungal Strain

White-rot, soft-rot and brown-rot fungi are the main groups of fungi involved in delignification. However, brown-rot and soft-rot fungi have a tendency to degrade cellulose and hemicellulose along with lignin [6]. Thus, white-rot fungi are the most common agent of delignification as it is the most selective group of fungi. They have the potential to remove lignin without severely degrading cellulose during the process. Most lignin-selective fungi from this group are, e.g., *P. chrysosporium, C. Subvermispora, Pleurotus ostreatus, Phlebia radiata, Phlebia tremellosus, Phlebia subserialis, Phellinus pini* and *Dichomitus squalens* [2]. Among these ligninolytic fungi, some are capable of producing the entire ligninolytic enzyme cocktail while others produce one or two of the ligninolytic enzymes. When grown on similar medium, *Pleurotus sajor-caju* and *Lentinus edodes* produced both MnP and laccase but not LiP whereas *P. chrysosporium* produced all the three enzymes when cultivated on a similar medium. This suggested that screening a large number of isolates to select the one with highest ligninolytic enzyme yield and delignification rate on a specific substrate is important [4].

5.5.2 Carbon Source and Concentration

The effect of various carbon sources on ligninolytic enzyme production has been studied, and the reports indicate that ligninolytic fungi exhibit a wide diversity in their response to ligninolytic enzyme production and lignin degradation in the presence of different carbon sources [82, 90, 91]. The supplementation of medium with easily metabolizable carbon is known to facilitate the production of fungal biomass and colonization in the deeper areas of lignocellulosic solid substrate. Among different carbon sources, glucose is the most commonly used carbon source for optimum production of fungal biomass and extracellular enzymes [74]. However, its effect on ligninolytic enzyme production is found to be strain specific as seen in a study where laccase activity of white-rot fungi *Pleurotus ostreatus* could not be enhanced on supplementation with glucose [92]. Various reports in the literature support that the use of additional carbon sources to supplement both synthetic media and lignocellulosic biomass is required for efficient fungal growth and enhanced production of ligninolytic enzymes [90–94].

5.5.3 Nitrogen Source and Concentration

There are numerous reports on the effect of nitrogen sources on ligninolytic enzyme production, and these studies indicated the nitrogen level in the culture medium is critical for the enzyme production and it varies widely among fungal strains and species. Both nitrogen-rich and nitrogen-deficient conditions can have stimulatory or inhibitory effect on ligninolytic enzyme production [90, 95]. However, nitrogen limitation is known to enhance lignin degradation. High nitrogen concentration is generally known to interfere with the selectivity of some species of white-rot fungi; hence, the removal of some nitrogen improves lignin degradation [96]. Additionally, different sources of nitrogen produce different responses in white-rot fungi. For instance, inorganic nitrogen source decreased the activities of laccase, MnP and peroxidases of *Pleurotus ostreatus* while organic nitrogen in the form of peptone and casein showed positive effects on these enzyme productions [95]. In another study, the different responses of nitrogen source and concentration were studied among various species and it was inferred that a low concentration of organic nitrogen source was preferred for enhanced ligninolytic enzyme production and lignin degradation [97]. For white-rot fungi, the depolymerization of lignin has mostly been observed to induce under nitrogen starvation [30, 98].

5.5.4 Aeration

Lignin degradation is an oxidative process, and thus air (oxygen) needs to be continually replenished to stimulate effective degradation. In flask culture method, passive diffusion of air through cotton plugs was found to be sufficient [99]. However, in large reactors containing packed biomass, active aeration is a primary requisite to provide uniform air distribution throughout the substrate. Aeration, in addition to oxygenation, is also required for removal of carbon dioxide, dissipation of heat, regulation of humidity, distribution of volatile compounds produced during fermentation [97]. A study on effect of oxygen on pretreatment of wheat straw by the *Pycnoporus cinnabarinus* and *Phanerochaete sordida* stated that aeration with oxygen three times a week reduced the reaction time by approximately one week [16]. Also, aeration at different flow rates for pretreatment of aspen chips by *P. chrysosporium* was tested and the flow rate of 0.022 vv^{-1} min^{-1} was found to be optimal for increased degradation rate [100]. Similar work by another researcher recorded improved production of LiP and MnP through aeration with pure oxygen [101]. Although high oxygen concentration could improve the rate of delignification, it does not have any effect on the delignification selectivity [99].

5.5.5 *Initial Moisture Content*

Initial moisture content is an essential factor during solid-state fermentation, and its significance on enzyme production and lignin degradation is depicted by the fact that it is essential for both substrate utilization and growth of any fungal strain and should be dined at the onset of substrate preparation. A lower solid-to-liquid ratio generally enhances MnP and LiP production [102]. The production of laccase was improved by increasing the moisture content from 40 to 60% [93]. Also, interaction of initial moisture content with other factors such as time significantly affected lignin degradation in cotton stalks by *P. Chrysosporium* [37]. Solid-state fermentation for pretreatment of corn cobs using *P. chrysosporium* was investigated in an initial moisture content range of 40–90%, and the highest lignin peroxidase activity was observed at 70% moisture content [103]. While studying the biological pretreatment of corn stover biomass by *C. subvermispora*, no fungal growth and lignin degradation were observed up to 45% initial moisture content while on increasing the initial moisture content to 60%, 75% and 85%, lignin degradation increased and reached up to 19.48%, 29.54% and 31.33%, respectively [40].

It is significant to provide an optimized initial moisture content and control the water activity of the fermenting substrate, since a higher or lower level adversely affects the physico-chemical properties of the solids in the substrate thus affecting the overall enzyme activity and delignification rate.

5.5.6 *pH*

Initial pH is one of the most important factors which markedly influence enzyme activity by altering the three-dimensional structure of enzymes. Active site amino acid residues of an enzyme often have acidic or basic properties, and any change from the optimal pH range can affect these residues and is difficult for substrates to bind. Thus, extremely high and low pH values outside optimal range may result in a complete loss of catalytic activity for most of the enzymes. Most fungi are known to grow best at a slightly acidic pH between 4 and 5. Also, some of these fungi while growing in the lignocellulosic feedstock also acidify them to a pH level that can partially or completely inhibit any further growth [30, 99, 104]. For example, wheat straw fermentation by *Dichomitus squalens* recorded a decrease in pH during growth [105]. Laccase production from *Pleurotus ostreatus* was observed to be optimal at pH 5.0, and the enzyme production decreased when pH was increased to more than 5.0 [93].

5.5.7 Temperature

The impact of incubation temperature on delignification rate and selectivity varies in different fungal strains since the optimal temperature for biological delignification of any lignocellulosic feedstock is dependent on the type and structural complexity of substrate, type of fungal strain and its physiology [106]. The fungal metabolism during the process of delignification generates heat which may raise the temperature of fermenting biomass to levels that are inhibitory for further growth and enzyme production. Therefore, heat dissipation is one of the significant factors to be considered for bioreactor design during scale-up study of solid-state fermentation. In general, white-rot Ascodiomycetes are moderate thermophiles and can grow significantly well at 39 ± 2 °C while white-rot Basidiomycetes are mesophiles with an optimum temperature range of 15–35 °C with better delignification rate observed between 25 and 30 °C [30, 99]. Lignin degradation rate when compared at 22 and 30 °C was found to be higher at 22 °C for *Ganoderma applanatum*, *Pleurotus ostreatus* and *Pleurotus serotinus* whereas at 30 °C for *Trametes hirsute* [104]. The effect of a wide range of reaction temperature (4–37 °C) during pretreatment of corn stover by *C. subvermispora* was evaluated, and 28 °C was concluded as the best temperature for enzymatic degradation of corn stover [40]. Lignin degradation from paddy straw by *Fusarium* sp. has also been reported at an optimum temperature of 27 ± 2 °C [107].

5.5.8 Substrate Particle Size

Substrate particle size is a major factor affecting the fungal growth, production of enzymes and lignin degradation especially in solid-state fermentation. Small particle size provides a larger surface area for microbial attack but may hinder respiration due to compaction and render ventilation ineffective resulting in poor performance of fungi. On the other side, a larger particle size provides better metabolic and ventilation efficiency because of the increased interparticle space. However, hindrance to the penetration of fungi and proper movement of air, water and metabolites to the substrate particles are some of the limitations of a large particle size. A study reported that a fine particle size of less than 0.42 mm resulted in lower delignification rate compared to 0.42–0.82 mm during pretreatment of rapeseed with P. *ostreatus* [108].

Substantially, a lower lignin degradation rate was achieved on using 15-mm particle size of corn stover compared to 5- and 10-mm particle size while no significant difference was observed between 5- and 10-mm particle sizes [40].

5.5.9 Incubation Time

Incubation period is one of the important parameters during biological pretreatment as long residence time owing to the low degradation rates is one of the major obstacles for industrial-scale application of biological pretreatment. A longer retention time promises a higher degradation percentage but a lower reaction rate with increasing time. Generally, several weeks to months are required for a high degree of degradation. On the contrary, a shorter residence time leads to a higher degradation rate per reactor volume unit, but overall low degradation percentage. Also, hemicellulose and cellulose along with lignin are also degraded due to non-selective nature of some fungi. As the lignin degradation occurs, the rate of carbohydrate degradation increases due to increased accessibility to enzymatic hydrolysis. Consequently, after an optimum incubation time no further increase in the yield of digestible polysaccharides occurs, even though lignin degradation continues [30, 104]. *P. chrysosporium*, a fast-growing white-rot fungus, only requires few days to weeks for effective delignification, but its non-selective nature is an undesirable trait for biofuel production [37, 58]. In contrast, *P. ostreatus* selectively resulted in 10.3% lignin degradation in beechwood in a few weeks [33]. Considerable lignin degradation during the fungal treatment of wheat bran by *P. ostreatus* occurred during the first 62 days [108]. For feedstocks such as softwood and bamboo which are resistant to fungal attack, even longer retention time is required. As reported in some studies, residence time of as long as 60-120 days was needed for treatment of bamboo with *Echinodontium taxodii* and *C. versicolor* to attain a significant degree of degradation [36].

References

1. Buswell JA, Odier E (1987) Lignin biodegradation. Critic Rev Biotechnol 6:1–60
2. Eriksson KEL, Blanchette RA, Ander P (1990) Microbial and enzymatic degradation of wood and wood components. Springer, New York Berlin Heidelberg
3. Zabel R, Morrell J (1992) Wood microbiology: decay and its prevention. Academic Press, London
4. Blanchette RA (1995) Degradation of lignocellulose complex in wood. Canad J Botany 73:S999–S1010
5. Martinez AT, Speranza M, Ruiz-Duenas FJ, Ferreira P, Camarero S, Guillen F et al (2005) Biodegradation of lignocellulosics: microbial, chemical, and enzymatic aspects of the fungal attack of lignin. Inter Microbiol 8:195–204
6. Sun Y, Cheng J (2002) Hydrolysis of lignocellulosic materials for ethanol production: a review. Bioresour Technol 83:111

7. Salvachua D, Prieto A, Lopez-Abelairas M, Lu-Chau T, Martínez AT, Martinez MJ (2011) Fungal pretreatment: an alternative in second-generation ethanol from wheat straw. Bioresour Technol 102:7500–7506
8. Nazarpour F, Abdullah DK, Abdullah N, Zamiri R (2013) Evaluation of biological pretreatment of rubberwood with white rot fungi for enzymatic hydrolysis. Materials 6:2059–2073
9. Camarero S, Garcoa O, Vidal T, Colom J, del Roo JC, Gutierrez A, Gras JM, Monje R, Marttonez MJ, Marttonez AT (2004) Efficient bleaching of non-wood high-quality paper pulp using laccase-mediator system. Enzyme Microb Technol 35:113–120
10. Tien K, Kirk TK, Farell RL (1987) Enzymatic combustion. Ann Rev Microbiol 41:465
11. Singh D, Zeng J, Laskar DD, Deobald L, Hiscox WC, Chen S (2011) Investigation of wheat straw biodegradation by *Phanerochaete chrysosporium*. Biomass Bioenergy 35:1030–1040
12. Zhang J, Ren X, Chen W, Bao J (2012) Biological pretreatment of corn stover by solid state fermentation of *Phanerochaete chrysosporium*. Front Chem Sci Eng 6(2):146–151
13. Atlas RM, Bartha R (1998) Biogeochemical cycling. Microbial ecology, vol 4. An imprint Addison Wesley Longman Inc., Sydney, pp 403–405
14. Lawson LR, Still CN (1957) The biological decomposition of lignin—a literature survey. Tappi J 40:56A–80A
15. Takahashi M (1976) Removal of lignin from partially delignified softwoods by soft rot and white rot fungi. Wood Res 61:1–10
16. Hatakka A (1994) Lignin-modifying enzymes from selected white-rot fungi: Production and role in lignin degradation. FEMS Microbiol Rev 13(3):125–135
17. Yu H, Guo G, Zhang X, Yan K, Xu C (2009) The effect of biological pretreatment with the selective white-rot fungus *Echinodontium taxodii* on enzymatic hydrolysis of softwoods and hardwoods. Bioresour Technol 100:5170–5175
18. Yao W, Nokes SE (2014) *Phanerochaete chrysosporium* pretreatment of biomass to enhance solvent production in subsequent bacterial solid-substrate cultivation. Biomass Bioenergy 62:100–107
19. Srinivasan C, D"Souza TM, Boominathan K, Reddy CA (1995) Demonstration of Laccase in the white rot basidiomycete *Phanerochaete chrysosporium* BKM-F1767. Appl Environ Microbiol 61(12):4274–4277
20. Glenn KJ, Gold HM (1983) Decolorization of several polymeric dyes by the lignin-degrading basidiomycete *Phanerochaete chrysosporium*. Appl Environ Microbiol 45:1741–1747
21. Bradley C, Wood P, Kearns R, Black B (1989) Biological delignification of wood and straw for ethanol production *via* solid state culture, Final Report. Montana Department of Natural Resources and Conservation, Montana
22. Kuhar S, Nair LM, Kuhad RC (2008) Pretreatment of lignocellulosic material with fungi capable of higher lignin degradation and lower carbohydrate degradation improves substrate acid hydrolysis and eventual conversion to ethanol. Canad J Microbiol 54:305–313
23. Jian S, Ratna R, Sharma-Shivappa Chinn M, Howell N (2008) Effect of microbial pretreatment on enzymatic hydrolysis and fermentation of cotton stalks for ethanol production. Biomass Bioenergy 33:88–96
24. Haider K, Trojanowski J (1980) A comparison of the degradation of 14C-labeled DHP and corn stalk lignins by micro-and macrofungi and bacteria. In: Kirk TK, Higuchi T, Chang H (eds) Lignin biodegradation: microbiology, chemistry and applications, vol 1. CRC Press Inc., USA, pp 111–134
25. Rodriguez A, Falcon MA, Carnicero A, Perestelo F, Fuente G, Trojanowski J (1996) Laccase activities of *Penicillium chrysog*enum in relation to lignin degradation. Appl Microbiol Biotechnol 45(3):399–403
26. Ferraz A, Baeza J, Duran N (1991) Softwood biodegradation by an ascomycete *Chrysonilia sitophila* (TFB 27441 strain). Lett Appl Microbiol 13:82–86
27. Liers C, Ullrich R, Steffen KT, Hatakka A, Hofrichter M (2006) Mineralization of 14C-labelled synthetic lignin and extracellular enzyme activities of the wood-colonizing ascomycetes *Xylaria hypoxylon* and *Xylaria polymorpha*. Appl Microbiol Biotechnol 69:573–579

28. Verkley GJM, Silva M, Wicklow DT, Crous PW (2004) *Paraconiothyrium*, a new genus to accommodate the mycoparasite *Coniothyrium minitans*, anamorphs of *Paraphaeosphaeria*, and four new species. Stud Mycol 50:323–335
29. Gao H, Wang Y, Zhang W, Wang W, Mu Z (2011) Isolation, identification and application in lignin degradation of an ascomycete GHJ-4. Afric J Biotechnol 10(20):4166–4174
30. Reid ID (1985) Biological delignification of aspen wood by solid-state fermentation with the white-rot fungus *Merulius tremellosus*. Appl Environ Microbiol 50:133–139
31. Regalado V, Rodriguez A, Perestelo F, Carnicero A, De La Fuente G, Falcon MA (1997) Lignin degradation and modification by the Soil-Inhabiting Fungus *Fusarium proliferatum*. Appl Environ Microbiol 63(9):3716–3718
32. Li X, Kondo R, Sakai K (2002) Biodegradation of sugarcane bagasse with marine fungus *Phlebia* sp. MG-60. J Wood Sci 48(2):159–162
33. Itoh H, Wada M, Honda Y, Kuwahara M, Watanabe T (2003) Bioorganosolve pretreatments for simultaneous saccharification and fermentation of beech wood by ethanolysis and white rot fungi. J Biotechnol 103:273–280
34. Songulashvili G, Elisashvili V, Penninckx M, Metreveli E, Hadar Y, Aladashvili N, Asatiani M (2005) Bioconversion of plant raw materials in value-added products by *Lentinus edodes* (Berk.) Singer and *Pleurotus* sp. Int J Med Mushrooms 7(3):467–468
35. Mahmoud YAG (2006) Biodegradation of water hyacinth by growing *Pleurotus ostreatus* and *P. sajor-caju* and trial for using in production of mushroom spawn. Acta Alimentaria 5(1):63–72
36. Zhang X, Yu H, Huang H, Liu Y (2007) Evaluation of biological pretreatment with white rot fungi for the enzymatic hydrolysis of bamboo culms. Int J Biodeterior Biodegradation 60:159–164
37. Shi J, Chinn MS, Sharma-Shivappa RR (2008) Microbial pretreatment of cotton stalks by solid state cultivation of *Phanerochaete chrysosporium*. Bioresour Technol 99:6556–6564
38. Singh P, Suman A, Tiwari P, Arya N, Gaur A, Shrivastava AK (2008) Biological pretreatment of sugarcane trash for its conversion to fermentable sugars. World Microbiol Biotechnol 24:667–673
39. Dias AA, Freitas GS, Marques GS, Sampaio A, Fraga IS, Rodrigues MA, Evtuguin DV, Bezerra RM (2010) Enzymatic saccharification of biologically pre-treated wheat straw with white-rot fungi. Bioresour Technol 101(15):6045–6050
40. Wan C, Li Y (2010) Microbial delignification of corn stover by *Ceriporiopsis subvermispora* for improving cellulose digestibility. Enzyme Microb Technol 47:31–36
41. Sari E, Syamsiah S, Sulistyo H, Hidayat M (2011) The kinetic of biodegradation lignin in water hyacinth (*Eichhornia Crassipes*) by *Phanerochaete Chrysosporium* using solid state fermentation (SSF) method for bioethanol production, Indonesia. Int J Chem Mole Nuc Mater Metallurg Eng 5(6):452–455
42. Nallapeta S, Nigam VK, Survajahala P, Mohan K (2012) Screening and selection of white rot fungi for biological delignification of agricultural residues. Int J Adv Biotechnol Res 3(4):790–796
43. Das S, Gangly A, Deya A, Ting Y-P, Chatterjee PK (2014) Characterization of water hyacinth biomass and microbial degradation of the biomass under solid state fermentation using a lignocellulolytic fungus (*Alterneria* sp. NITDS1). J Chem Biolog Phys Sci 4(3):2279–2293
44. Sari E, Syamsiah S, Sulistyo H, Hidayat M (2015) Effect of $Mn_2 +$ addition on delignification of water hyacinth using *Phanerochaete chrysosporium*. Modern Appl Sci 9(2):228
45. Machado ADS, Ferraz A (2017) Biological pretreatment of sugarcane bagasse with basidiomycetes producing varied patterns of biodegradation. Bioresour Technol 225:17
46. Manickam NK, Rajarathinam R, Muthuvelu KS, Senniyappan T (2018) New insight into the effect of fungal mycelia present in the bio-pretreated paddy straw on their enzymatic saccharification and optimization of process parameters. Bioresour Technol 267:291
47. Ball AS, Betts WB, McCarthy AG (1989) Degradation of lignin-related compounds by actinomycetes. Appl Environ Microbiol 55:1642–1646

48. Colberg PJ (1988) Anaerobic microbial degradation of cellulose, lignin, oligolignols, and monoaromatic lignin derivatives. In: Biology of anaerobic microorganisms. Wiley, New York USA, pp 333–372
49. Morii H, Nakamiya K, Kinoshita S (1995) Isolation of lignin decoloursing bacterium. J Ferm Bioeng 80:296–299
50. Perestelo F, Rodriquez A, Perez R, Carnicero A, Fuente G, Falcon MA (1996) Isolation of a bacterium capable of limited degradation of industrial and labeled natural and synthetic lignins. World J Microbiol Biotechnol 12:111–112
51. Abd-Elsalam HE, El-Hanafy AA (2009) Lignin biodegradation with ligninolytic bacterial strain and comparison of *Bacillus subtilis* and *Bacillus* sp. isolated from Egyptian Soil. American-Eurasian J Agri Environ Sci 5(1):39–44
52. Amin FR, Khalid H, Zhang H, Rahman S, Zhang R, Liu G, Chen C (2017) Pretreatment methods of lignocellulosic biomass for anaerobic digestion. AMB Expr 7:72
53. Ramachandra M, Crawford DL, Hertel G (1988) Characterization of an extracellular lignin peroxidase of the lignocellulolytic actinomycete *Streptomyces viridosporus*. Appl Environ Microbiol 54:3057–3063
54. Chandra R, Raj A, Purohit HJ, Kapley A (2007) Characterisation and optimisation of three potential aerobic bacterial strains for kraft lignin degradation from pulp paper waste. Chemosphere 67:839–846
55. Abd-Elsalam HE, El-Hanafy AA (2009) Lignin biodegradation with ligninolytic bacterial strain and comparison of *Bacillus subtilis* and *Bacillus* sp. isolated from Egyptian soil. American-Eurasian J Agri Environ Sci 5:39–44
56. Bandounas L, Wierckx NJP, Winde JH, Ruijssenaars HJ (2011) Isolation and characterization of novel bacterial strains exhibiting ligninolytic potential. BMC Biotechnol 11:94
57. Yang YS, Zhou JT, Lu H, Yuan YL, Zhao LH (2012) Isolation and characterization of *Streptomyces* sp. strains F-6 and F-7 capable of decomposing alkali lignin. Environ Technol 33:2603–2609
58. Shi Y, Chai L, Tang C, Yang Z, Zheng Y, Chen Y, Jing Q (2013) Biochemical investigation of kraft lignin degradation by *Pandoraea* sp. B-6 isolated from bamboo slips. Bioproc Biosys Eng 36:1957–1965
59. Chandra R, Bharagava RN (2013) Bacterial degradation of synthetic and kraft lignin by axenic and mixed culture and their metabolic products. J Environ Biology 34:991–999
60. Chang YC, Choi D, Takamizawa K, Kikuchi S (2014) Isolation of *Bacillus* sp. strains capable of decomposing alkali lignin and their application in combination with lactic acid bacteria for enhancing cellulase performance. Bioresour Technol 152:429–436
61. Lu C, Wang H, Luo Y, Guo L (2010) An efficient system for pre-delignification of gramineous biofuel feedstock in vitro: application of a laccase from *Pycnoporus sanguineus* H275. Process Biochem 45:1141–1147
62. Rolz C, Leon R, Arriola MC, Cabrera S (1986) Biodelignification of lemon grass and citronella bagasse by white rot fungi. Appl Environ Microbiol 52(4):607–611
63. Bhandari KS, Bist V (1989) Effect of *Coriolus versicolor* on physicochemical properties of *Eucalyptus globules* wood. Wood Sci Technol 23:163–169
64. Tuomela M, Oivanen P, Hatakka A (2002) Degradation of synthetic [^{14}C] lignin by various white-rot fungi in soil. Soil Biology Biochem 34(11):1613–1620
65. Luna ML, Murace MA, Keil GD, Otano ME (2004) Patterns of decay caused by *Pycnoporus sanguineus* and *Ganoderma lucidum* (Aphyllophorales) in poplar wood. IAWA Journal 25(4):425–433
66. Breccia JD, Sineriz F, Bettucci L, Piaggio M (2004) Degradation of sugarcane by several white rot fungi. Acta Biotechnol 17:177–184
67. Tapia C, Jimenez I, Medina J, Gonzalez J, Carmona R (2005) Evaluation of the lignolytic effect of the white rot fungi *Ceriporiopsis* sp., *Pleurotus* sp. and *Phlebia* sp. on industrial Pinus radiate logs. J Wood Chem Technol 25(1–2):81–94
68. Zeng GM, Shi JG, Yuan XZ, Liu J, Zhang ZB, Huang GH, Li JB, Xi BD, Liu HL (2006) Effects of tween 80 and rhamnolipid on the extracellular enzymes of *Penicillium simplicissimum* isolated from compost. Enzyme Microb Technol 39:1451–1456

69. Beek TAV, Kuster B, Claassen FW, Tienvieri T, Beataud F (2007) Fungal biotreatment of spruce wood with *Trametes versicolor* for pitch control: Influence on extractive contents, pulping process parameters, paper quality and effluent toxicity. Bioresour Technol 98(2):302–311
70. Gupta R, Mehta G, Khasa YP, Kuhad RC (2011) Fungal delignification of lignocellulosic biomass improves the saccharification of cellulosics. Biodegradation 22:797–804
71. Geethanjali PA (2012) A study on lignin degrading fungi isolated from the litter of evergreen forests of Kodagu, Karnataka. Int J Environ Sci 2(4):2034–2039
72. Mukherjee R, Nandi B (2004) Improvement of in vitro digestibility through biological treatment of water hyacinth biomass by two *Pleurotus* sp. Int Biodeterior Biodegradation 53(1):7–12
73. Ferraz A, Cordova AM, Machuca A (2003) Wood biodegradation and enzyme production by *Ceriporiopsis subvermispora* during solid state fermentation of *Eucalyptus grandis*. Enzyme Microb Technol 32:59–65
74. Sharma RK, Arora DS (2010) Production of lignocellulosic enzymes and enhancement of invitro digestibility during solid state fermentation of wheat straw by *Phlebia floridensis*. Bioresour Technol 101:9248–9253
75. Xu C, Ma F, Zhang X (2009) Lignocellulose degradation and enzyme production by *Irpex lacteus* CD2 during solid-state fermentation of corn stover. J Biosci Bioeng 108:372–375
76. Khuong LD, Kondo R, Leon RD, Anh TK, Meguro S, Shimizu K, Kamei I (2014) Effect of chemical factors on integrated fungal fermentation of sugarcane bagasse for ethanol production by a white-rot fungus, *Phlebia* sp. MG-60. Bioresour Technol 167:33–40
77. Maijala P (2005) Co-culturing of white rot fungi on wood—potential in biopulping. In: Seminar cum workshop on forest pathology
78. Chi Y, Hatakka A, Maijala P (2007) Can co-culturing of two white rot fungi increase lignin degradation and the production of lignin degrading enzyme. Int Biodeterior Biodegradation 59(1):32–39
79. Dwivedi P, Vivekanand V, Pareek N, Sharma A, Singh RP (2010) Bleach enhancement of mixed wood pulp by xylanase-laccase concoction derived through co-culture strategy. Appl Biochem Biotechnol 160(1):255–268
80. Kuila A, Mukhopadhyay M, Tuli DK, Banerjee R (2011) Accessibility of enzymatically delignified Bambusa bambos for efficient hydrolysis at minimum cellulose loading: an optimization study. Enzyme Res 1–8
81. Moniruzzaman M, Ono T (2012) Ionic liquid assisted enzymatic delignification of wood biomass: a new “green” and efficient approach for isolating of cellulose fibers. Biochem Eng J 60:156–160
82. Wang F, Xie H, Chen W, Wang E, Du F, Song A (2013) Biological pretreatment of corn stover with ligninolytic enzyme for high efficient enzymatic hydrolysis. Bioresour Technol 144:572–578
83. Babot ED, Rico A, Rencoret J, Kalum L, Lund H, Romero J, Del Rio JC, Martinez AT, Gutierrez A (2011) Towards industrially-feasible delignification and pitch removal by treating paper pulp with *Myceliophthora thermophila* laccase and a phenolic mediator. Bioresour Technol 102:6717–6722
84. Chen Q, Marshall MN, Geib SM, Tien M, Richard TL (2012) Effects of laccase on lignin depolymerization and enzymatic hydrolysis of ensiled corn stover. Bioresour Technol 117:186–192
85. Qiu W, Chen H (2012) Enhanced the enzymatic hydrolysis efficiency of wheat straw after combined steam explosion and laccase pretreatment. Bioresour Technol 118:8–12
86. Placido J, Sergio Capareda S (2014) Analysis of alkali ultrasonication pretreatment in bioethanol production from cotton gin trash using FT-IR spectroscopy and principal component analysis. Bioresour Bioprocess 1:23
87. Kamei I, Hirota Y, Meguro S (2012) Integrated delignification and simultaneous saccharification and fermentation of hard wood by a white-rot fungus, *Phlebia* sp. MG-60. Bioresour Technol 126:137–141

88. Ma K, Ruan Z (2015) Production of a lignocellulolytic enzyme system for simultaneous bio-delignification and saccharification of corn stover employing co-culture of fungi. Bioresour Technol 175:586–593
89. Mate DM, Alcalde M (2015) Laccase engineering: from rational design to directed evolution. Biotechnol Adv 33:25–40
90. Elisashvili V, Penninckx M, Kachlishvili E, Asatiani M, Kvesitadze G (2006) Use of *Pleurotus dryinus* for lignocellulolytic enzymes production in submerged fermentation of mandarin peels and tree leaves. Enzyme Microb Technol 38:998–1004
91. Kaal EEJ, Field JA, Joyce TW (1995) Increasing ligninolytic enzymatic activities in several white rot basidiomycetes by nutrient sufficient media. Bioresour Technol 59:133
92. Ardon O, Kerem Zohar, Hadar Y (1996) Enhacement of laccase activity in liquid cultures of the lignolytic fungus *Pleurotus ostreatus* by cotton stalk extract. J Biotechnol 51:201–207
93. Patel H, Gupte A, Gupte S (2009) Effect of different culture conditions and inducers on production of laccase by a Basidiomycete fungal isolate *Pleurotus ostreatus* HP-1 under solid state fermentation. BioResources 4(1):268–284
94. Zhang Z, Xia L, Wang F, Lv P, Zhu M, Li J, Chen K (2015) Lignin degradation in corn stalk by combined method of H_2O_2 hydrolysis and *Aspergillus oryzae* CGMCC5992 liquid-state fermentation. Biotechnol Biofuels 8:183
95. Mikiashvili N, Wasser SP, Nevo E, Elisashvili V (2006) Effects of carbon and nitrogen sources on *Pleurotus ostreatus* ligninolytic enzyme activity. World J Microbiol Biotechnol 22:999–1002
96. Dorado J, Field J, Almendros G, Sierra-Alvarez R (2001) Nitrogen-removal with protease as a method to improve the selective delignification of hemp stemwood by the white-rot fungus *Bjerkandera* sp. strain BOS55. Appl Microbiol Biotechnol 57:205–211
97. Isroi I, Millati R, Syamsiah S et al (2011) Biological treatment of lignocelluloses with white-rot fungi and its applications: a review. BioResources 6:5224–5259
98. Ruttimann-Johnson C, Salas L, Vicuna R, Kirk TK (1993) Extracellular enzyme production and synthetic lignin mineralization by *Ceriporiopsis subvermispora*. Appl Environ Microbiol 56:1792–1797
99. Reid ID (1989) Solid-state fermentations for biological delignication. Enzyme Microb Technol 11:786–803
100. Messner K, Koller K, Wall MB, Akhtar M, Scott GM (1998) Fungal treatment or wood chips for chemical pulping. In: Environmental friendly technologies for the pulp and paper industry. Wiley, Inc. New York, pp 385–419
101. Belinky PA, Flikshtein N, Lechenko S, Gepstein S, Dosoretz CG (2003) Reactive oxygen species and induction of lignin peroxidase in *Phanerochaete chrysosporium*. Appl Environ Microbiol 69:6500–6506
102. Fujian X, Hongzhang C, Zuohu L (2001) Solid-state production of lignin peroxidase (LiP) and manganese peroxidase (MnP) by *Phanerochaete chrysosporium* using steam-exploded straw as substrate. Bioresour Technol 80:149–151
103. Asgher M, Ahmad Z, Iqbal HMN (2013) Alkali and enzymatic delignification of sugarcane bagasse to expose cellulose polymers for saccharification and bio-ethanol production. Ind Crop Prod 44:488–495
104. Zadrazil F, Brunnert H (1982) Solid state fermentation of lignocellulose containing plant residues with *Sporotrichum pulmerulentum Nov.* and *Dichomitus squalens (Karst.) reid.* Eur J Appl Microbiol Biotechnol 16:45–51
105. Agosin E, Daudin JJ, Odier E (1985) Screening of white-rot fungi on (^{14}C) lignin-labelled and (^{14}C) whole-labelled wheat straw. Appl Microbiol Biotechnol 22:132–138
106. Levonen-Munoz E, Bone DH, Daugulis AJ (1983) Solid state fermentation and fractionation of oat straw by basidiomycetes. Eur J Appl Microbiol Biotechnol 18:120
107. Sahni N, Phutela UG (2013) Comparative profile of paddy straw pretreated with standard and isolated lignocellulolytic fungal cultures. J Yeast Fungal Res 4(7):92–97
108. Locci E, Laconi S, Pompei R, Scano P, Lai A, Marincola FC (2008) Wheat bran biodegradation by *Pleurotus ostreatus*: a solid-state Carbon-13 NMR study. Bioresour Technol 99:4279–4284

Chapter 6
Strategies for Saccharification of Lignocellulosic Substrate

Abstract Lignocellulosic materials, such as agricultural residues, softwood, hardwood, agro-industrial and forest-industrial residues are abundant renewable resources of biofuels to replace fossil fuels. The conversion of lignocellulosic biomass to bioethanol occurs in three main stages, an effective pretreatment step for deconstruction of biomass followed by enzymatic hydrolysis to produce fermentable sugars and finally fermentation of these sugars to ethanol. Enzymatic hydrolysis is one of the most essential steps in the bioethanol production and involves the conversion of cellulose and hemicellulose to simple sugars using enzymatic complex consisting of cellulases and hemicellulases. There are various limitations associated with enzymatic hydrolysis that hinders with the economic feasibility of the process. Various strategies to overcome these limitations include an efficient pretreatment method, optimization of various factors, recycling of the enzymes and utilization of high solid concentrations. The current chapter ponders upon the mechanism of enzymatic hydrolysis, enzymes associated with the process, various strategies like SSF, SmF and analytical saccharification and various factors affecting the process of enzymatic saccharification.

Keywords Cellulase · Fermentation · Hydrolysis · Saccharification · Trichoderma · Xylose

The lignocellulosic biomass is the largest source of carbohydrate and is thus, an important and sustainable resource for bio-based fuels, chemicals and materials. Lignocellulosic feedstocks include energy crops, agricultural residues like wheat straw, rice straw, corn stover, sugarcane bagasse, etc., softwood, hardwood, agro-industrial residues and industrial waste. It consists of lignin chains, cellulose and hemicellulose intertwined together by non-covalent forces and covalent crossed connections, making it resistant to enzymatic hydrolysis. In order to be used as a resource, any lignocellulosic biomass has to be first deconstructed and converted to simpler sugars that can further be utilized and converted to biofuels and other valuable bio-based products. This deconstruction of biomass to sugars requires biochemical conversion with enzymatic hydrolysis of cellulose using various enzymes as the core process [1]. However, biochemical conversion of lignocelluloses is hindered by their extreme inherent recalcitrance to enzymatic hydrolysis. The greatest obstacle in the use of the

A. Sharma and N. K. Aggarwal, *Water Hyacinth: A Potential Lignocellulosic Biomass for Bioethanol*, https://doi.org/10.1007/978-3-030-35632-3_6

lignocellulosic residues is represented by the structural barrier formed by the lignin due to its syringyl and guaiacyl units and aryl ether linkages, and strong associations with hemicellulose which prevents the enzymes from penetrating and retards enzymatic conversion of cellulose. The removal and separation of the lignin may be achieved through various methods like physical, chemical or biological pretreatments or their combination. The most common of these is the chemical treatment which is usually accomplished through acid or alkaline hydrolysis [2].

6.1 Enzymatic Saccharification of Lignocellulosic Feedstocks

Enzymatic saccharification of cellulose is carried out by cellulases, a mixture of a large number of different cellulose hydrolysing enzymes that act in synergy to hydrolyse both crystalline and amorphous domains of cellulose to fermentable sugars [3]. The most studied cellulolytic and hemicellulolytic enzymes producing microorganism is the ascomycete *Trichoderma reesei* [4]. The hydrolysis of cellulose by cellulolytic enzymes has been investigated extensively and a great amount of efforts have been invested with the sole objective of developing an efficient process for biofuel production [5, 6]. Enzymes that are commonly responsible for the degradation of cellulose include the hydrolases β-endoglucanase (endo-1,4-b-D-glucanase, EC 3.2.1.4), β-exoglucanase (1,4-b-D-glucan-cellobiohydrolase, EC 3.2.1.91) or cellobiohydrolases and β-D-glucosidase (β-D-glucosideglucanohydrolase, EC 3.2.1.21) or cellobiases [3]. Endoglucanases randomly cleave the internal glycosidic bonds of cellulose chains along the amorphous regions producing reducing or non-reducing ends of cellooligosaccharides; cellobiohydrolases then hydrolyse the crystalline regions of the cellulose chain from the reducing or non-reducing end yielding cellobiose as major product and β-glucosidases then catalyse the cleavage of cellobiose to produce glucose. Other enzymes like lytic polysaccharide monooxygenase (LPMO) provide assistance to these major cellulolytic enzymes by disrupting crystalline cellulose in a reaction driven by molecular oxygen making cellulose more accessible to them [7, 8]. Since xylans are quantitatively the most important hemicelluloses present in agricultural residues and hardwood, most of the enzyme preparations used for enzymatic hydrolysis of these lignocelluloses also contains enzymes like xylanases, xylosidases and acetyl esterases. Additionally, to boost the enzymatic hydrolysis in such feedstocks, xylanase-rich enzyme preparations can be added separately along with cellulases.

Enzymatic hydrolysis is the most complex step in the process of conversion of lignocellulosic biomass to bioethanol due to substrate–enzyme interactions and their effects. The enzymatic saccharification is usually conducted at a temperature of 45–50 °C for 24−72 h with agitation [9, 10]. The enzymatic reactions may occur under mild conditions of pH 4–6 and does not have a corrosion problem. Additives such as antimicrobial agents are added to prevent loss of sugar from consumption by bacterial contaminants and surfactants such as Tween 20 to improve the enzymatic saccharification of cellulose are also added in the reaction mixture [11]. Enzymatic saccharification of biomass can be achieved either by crude enzyme extract or commercial enzyme preparations. Commercial cellulase preparations are mainly obtained from aerobic cultures of *Trichoderma reesei* and *Aspergillus niger* [12, 13]. One of the most commonly used commercial cellulolytic enzymes, Novozymes' Celluclast 1.5 L, is derived from *T. reesei*. However, it exhibits relatively low β-glucosidase activity and is often needed to be used or supplemented with a β-glucosidase-rich enzyme preparation, Novozym 188 obtained from *Aspergillus niger* [14, 15]. However, more recently developed enzyme preparations typically contain sufficient quantities of β-glucosidase activity and can be used without supplementation [15]. The major industries involved in industrial enzyme production are Novo Industri A/S (Denmark), Danisco/Genencor (Denmark and USA), BASF (Germany) and DSM (Netherlands) [16].

Both bacteria and fungi are known to produce cellulolytic enzymes responsible for the hydrolysis of lignocellulosic biomass. Bacterial species belonging to genera *Bacillus*, *Bacteriodes*, *Clostridium*, *Cellulomonas*, *Erwinia*, *Thermomonospora*, *Ruminococcus*, *Acetovibrio*, *Microbispora* and *Streptomyces* are widely known to produce cellulases. *Cellulomonas fimi*, *Clostridium thermocellum*, *Thermomonospora fusca* and *Bacillus subtilis* stand out among these cellulase-producing bacteria and have been studied extensively [5, 17, 18]. Thermophilic bacteria, *Geobacillus stearothermophilus*, has been provided special attention for the production of cellulases using separate saccharification and fermentation process of lignocellulosic biomass due to their higher temperature requirement and broad substrate range [19]. The most studied species of cellulase-producing fungi are *Trichoderma viride*, *Aspergillus niger*, *Penicillium funiculosum* and *Rhizopus oligosporus* [20, 21]. Other fungi that have been reported to produce cellulolytic enzymes include *P. chrysosporium*, *Sclerotium rolfsii* and other species of *Trichoderma*, *Aspergillus*, *Fusarium*, *Schizophyllum* and *Penicillium* [5, 18, 22]. These micro-organisms have been reported to grow on various lignocellulosic substrates for the production of cellulolytic enzymes under different cultivation conditions [2, 18, 22]. *Trichoderma* sp. has demonstrated capacity to produce the cellulolytic enzymes from various substrates like corncob, rice straw, wheat straw and sugarcane straw [21].

The operating cost of enzymatic hydrolysis is lower compared to acid or alkaline hydrolysis methods. The chemical hydrolysis presents advantages such as high conversion rate but the enzymatic hydrolysis is superior to it as it can be performed at low temperatures and atmospheric pressure and there is no toxic products formation [5]. Enzymatic hydrolysis also offers the potential for higher sugar yields and higher selectivity than chemical processes but the detailed mechanism of enzymatic hydrolysis and the relationship between the complex structure of substrate and function of various enzymatic components is not fully understood. Also, to reach high conversion rate of cellulose, high concentrations of enzymes is required. So, high cost of commercial enzyme preparations is another limitation of enzymatic hydrolysis [20]. The process of enzymatic hydrolysis can be performed in both solid-state and submerged culture media. SSF has several economical and practical advantages over SmF as it is cheaper, less technology oriented; thus, low energy requirements and the enzyme extraction and downstream processing are easier with limited solvent requirement and lower recovery cost than SmF [23]. Other significant benefits of using SSF over SmF are use of raw materials as substrates, lower wastewater output in the form of liquid effluent, potential higher volumetric productivity and high reproducibility. Since the moisture level requirement of the SSF is lower, a higher volumetric substrate loading is possible in SSF resulting in compact fermentation facility, simpler fermentation media and easier control of contamination [24]. Also, studies have indicated higher titres and better stability of secreted enzymes and a lower level of catabolic repression in SSF than in SmF while using the same strain, substrate and fermentation media [25, 26].

6.2 Solid-State Fermentation (SSF)

Solid-state fermentation is the process of cultivation of micro-organisms on solid substrate in the form of inert carriers or insoluble solid substrates that can also be used as a source of carbon and energy needed for microbial growth. The fermentation takes place in the presence of enough moisture to guarantee the cellular growth and metabolism but does not exceed the maximum retention capacity of the solid matrix, thus providing the closest possible natural environment in which the micro-organisms are normally adapted to grow [27]. The substrates used in solid-state fermentation are usually agro-industrial residues with low or no commercial value [28]. The hyphal growth and high tolerance to high osmotic pressure and low water activity (A_W) give fungi major advantages over bacteria for growth on solid substrates [29]. An efficient pretreatment step before enzymatic hydrolysis has also been known to enhance the process of saccharification. However, pretreatment is found to be more efficient in lignocellulosic materials where the accessibility to cellulose is hindered. The efficiency of the cellulolytic enzymes may also be affected because of the heterogeneity of lignocellulolytic substrates since they vary widely in the amounts of cellulose, hemicellulose and lignin in their biomass composition [18].

Pretreated corn stover was subjected to enzymatic hydrolysis using Celluclast 1.5L, a commercial *Trichoderma reesei* cellulase preparation and its ability to hydrolyse the cellulose and xylan components was significantly improved by supplementation with three other crude commercial enzyme preparations containing xylanase, pectinase and β-glucosidase. β-glucosidase contributed to relief from product inhibition while other accessory enzymes such as xylanase and pectinase hydrolysed non-cellulosic polysaccharides that coat cellulose fibres, thus improving the overall cellulase performance and cellulose hydrolysis. This enzyme mixture resulted in a 99 and 88% conversion of glucan to glucose and xylan to xylose, respectively [30]. Enzymatic saccharification of cotton by cellulase was done after pretreatment using cellulose dissolution method. Four different agents, NaOH/Urea solution, *N*-methylmorpholine-*N*-oxide (NMMO), ionic liquid (1-butyl-3-methylimidazolium chloride; [BMIM]Cl) and 85% phosphoric acid were used and the dissolution process was operated under a mild condition of temperature <130 °C and atmospheric pressure. The pretreated substrate exhibited a significant enhancement in saccharification rate with 87–96% saccharification yield as compared to only around 23% for the untreated substrate after 72 h. Cellulose regenerated from dissolution pretreatment with phosphoric acid achieved the highest saccharification rates and yield [31]. In another study, crude cellulase and a relatively glucose tolerant β-glucosidase produced from *Trichoderma reesei* RUT C30 and *Aspergillus niger* MTCC 7956, respectively, under solid-state fermentation with wheat bran as the substrate were employed for enzymatic hydrolysis of sugarcane bagasse, rice straw and water hyacinth biomass. Saccharification of these feedstocks was carried out using 50 FPU of cellulase and 10 U of β-glucosidase per gram of pretreated biomass and the maximum amount of reducing sugars (26.3 g/L) was produced from rice straw followed by sugarcane bagasse [20]. Three different samples including sugarcane bagasse pulp and skin and microcrystalline cellulose (Avicel) were pretreated with SC–CO_2 explosion at 12, 14 and 16 MPa at a temperature of 60 °C before enzymatic hydrolysis. The enzymatic hydrolysis reaction was carried at 55 °C for 8 h using cellulase. The glucose yield was found to be higher in the pretreated sample and was further enhanced when combined with alkali pretreatment [32]. Alkali-pretreated water hyacinth was used as a substrate for cellulase production by *A. niger* under solid-state fermentation (SSF) and the crude enzyme cocktail containing cellulase (9.8 FPU/gram dry substrate), endoglucanase (180 U/gds), β-glucosidase (178 IU/gds) and xylanase (260 IU/gds) was used for enzymatic hydrolysis. The highest reducing sugar concentration obtained was 13.5 g/L after 72 h [33]. Enzymatic saccharification of pretreated rice straw was carried out using cellulase produced from *Bacillus carboniphilus* CAS 3 and the optimal value of temperature, pH and NaCl were determined as 50 °C, 9 and 30%, respectively. Also, more than 70% of the original enzyme activity was retained even at extreme conditions of 80 °C, 12 pH and 35% NaCl suggesting it to have potential in harsh industrial and biotechnological applications [34]. Enzymatic saccharification of filter paper, eucalyptus sawdust and corncob was carried out using culture of *Trichoderma viride* and the cellulolytic activity was measured and compared with that of commercial cellulase enzyme. When comparing both forms of cellulase enzymes used, the cellulolytic complex obtained from *Trichoderma viride* showed lower enzyme activities during saccharification of the filter paper and the corncob while higher activities during saccharification of the eucalyptus sawdust. Additionally, both cellulolytic complex from *Trichoderma viride* and the commercial cellulase showed better activity and stability between temperature range of 40 and 50 °C [21].

6.3 Submerged Fermentation

Enzymatic hydrolysis of cellulose to glucose in submerged fermentation often occurs at slower reaction rates due to the highly complex and crystalline structure of cellulose and limited accessibility of enzyme adsorption sites in aqueous media. In a study, the cellulose structure (Avicel PH-101) was disrupted using the ionic liquid, 1-*n*-butyl-3-methylimidazolium chloride to accelerate the subsequent hydrolysis reaction. ILs being extremely non-volatile in nature dissolves cellulose without showing any negative impact on air quality compared to volatile solvent systems. Also, the enzymatic hydrolysis rate was nearly 50-fold higher for regenerated cellulose when compared to the untreated one [35]. In another study, enzymatic saccharification of untreated and alkali-treated date palm waste was performed by cellulase enzyme derived from *Geobacillus stearothermophilus*. The maximum saccharification rate of 71.03% with a yield of 31.56 mg/mL of glucose was obtained at a substrate and enzyme concentration of 4% and 30 FPU/g of substrate, respectively, after 24 h of incubation. The optimum pH and temperature for the bioconversion process were found to be 5.0 and 50 °C. The saccharification rate was further increased to 94.88% by removal of the hydrolysate after 24 h using a two-step hydrolysis process [19]. Another study on enzymatic saccharification reported that the pretreatment with 2% NaOH enhanced the enzymatic hydrolysis of wheat straw, rice straw and bagasse by *Bacillus subtilis* cellulases with saccharification rates of 33.0%, 25.5% and 35.5%, respectively [36].

6.4 Analytical-Scale Enzymatic Saccharification

Analytical enzymatic saccharification is a technique based on the screening of large sets of lignocellulosic feedstocks samples for biofuel production. Analytical-scale studies of enzymatic saccharification utilize a small amount of biomass samples to screen a large number of samples and parameters. Various parameters tested during this process may involve comparison of different lignocellulosic feedstocks, different parts of plants, different pretreatment techniques, and various environmental and cultural factors. Several different approaches used in analytical-scale enzymatic saccharification with or without pretreatment are semi-automated, high-throughput (HTP), microtiter plate (MTP) and microwell plates (MWP) based methods [1]. Some of these approaches also involve additional steps like diminution of the solid substrate before pretreatment and analysis of the saccharification or fermentation products. The selection of approach generally depends on the type and structural complexity of lignocellulosic material, the quantity of biomass available for analysis, the number of biomass samples and the number of variables being examined. Reduction in particle size of the solid substrate using grinding or diminution followed by a size fractionation step is usually done before pretreatment to facilitate uniform dispersion of biomass in vials or microplates. As the surface area of substrate particle has a huge impact on the rate of enzymatic saccharification, use of a well-defined particle size usually in the range 0.1–1.0 mm (20–80 mesh size) is recommended [37]. Also, proper drying and analysis of moisture content of the substrate is essential as yield calculations are typically based on the dry weight.

Pretreatment technique used for analytical enzymatic saccharification is based on both pressurized and non-pressurized systems. Examples of pressurized systems include autoclaves, steam-chamber reactor, solvent extractor, parr reactors, single-mode microwave systems and automated 96-well plate robotic systems, while the non-pressurized systems include water baths and heat blocks [1]. Most of the analytical-scale saccharification studies have been based on HPT using steam or dilute acid, or alkali in pressurized chambers. For instance, pretreatment using sodium hydroxide and ionic liquids has been performed in non-pressurized systems [38, 39]. Alkali and dilute acid pretreatment has also been used for aqueous pretreatment systems at a temperature and reaction time range from 80 to 200 °C and 3 min to 24 h, respectively. The process of separation of the solid and liquid phases can be done by centrifugation followed by thorough washing and neutralization with water and buffer or just neutralization using concentrated buffer. Separation of the liquid and solid phases prior to enzymatic hydrolysis is required for the ionic strength, removal of inhibitory compounds and separate analysis of sugar produced pretreatment and enzymatic saccharification [38, 40, 41]. Enzymatic saccharification of hydrolysate is usually performed using individual or combinations of commercial enzymes like cellulase preparations, such as Celluclast 1.5, Spezyme CP, Accelerase 1000 and Cellic CTec2, hemicellulase preparations such as Multifect xylanase, all obtained from *Trichoderma reesei*, β-glucosidase preparations, such as Novozym 188 obtained from *A. niger*. The samples obtained after are analysed using various methods including spectrophotometric technique for reducing sugars and enzymes assay, mass spectrometry (MS), high-performance liquid or anion-exchange chromatography (HPLC and HPAEC) and capillary electrophoresis (CE). Analytical enzymatic saccharification is very useful for screening a large numbers of samples, detailed analysis of reaction mixtures and to study the role of various factors contributing to the recalcitrance of lignocelluloses.

6.5 Factors Affecting Enzymatic Hydrolysis of Cellulose

Various nutritional and fermentation parameters affect enzyme production both under submerged as well as solid-state conditions. The fermentation medium must meet the nutritional requirements of the micro-organism and thus need to be optimized for maximum enzyme production. The efficiency of the enzymatic saccharification depends on several factors such as reaction time, temperature, pH, sources of carbon, nitrogen and some growth factors, the type of pretreatment of the substrate, enzymes and substrate concentrations, and catalytic action of cellulolytic enzyme complex.

6.5.1 Substrate Type and Concentration

The major constraint in the process of enzymatic saccharification is the cost of cellulase enzymes. However, production cost of cellulases may be controlled by the use of cheap lignocellulosic substrates like agro-industrial residues for production of the enzymes using solid-state fermentation (SSF) as the cost-effective fermentation technique. Several researchers have reported that agro-industrial residues such as rice bran, rice straw, wheat bran, wheat straw, sorghum straw, sugarcane bagasse, corn cob, cotton flower and cassava residue could be used as substrates for cellulase production [42–44]. *Bacillus subtilis* CBTK 106, *Bacillus subtilis* BL62 and *Bacillus pumillus* were found to exhibit their maximum cellulase productivity on supplementation of production medium with wheat bran, banaba stalk and soyabean, respectively [45, 46]. Higher cellulase titres were obtained by *Aspergillus* sp. on using different ratios of rice straw and wheat bran as feedstocks [47]. Five prospective *Aspergillus* sp. for cellulase production were isolated and the highest production of endo-β-glucanase was reported from *Aspergillus* MAM-F23 [48]. In another study, *A. niger* was cultivated under submerged fermentation with sorghum straw and wheat straw as substrate for cellulase production and maximum enzyme yield was reported from sorghum straw [49].

An increased substrate concentration would lead to an increase in reaction rate and glucose yield while a low substrate concentration would result in an overall loss in glucose yield. However, a high substrate concentration would result in substrate inhibition eventually leading to a decrease in the rate of the hydrolysis. The extent of substrate inhibition during the process of enzymatic saccharification depends on the substrate to enzyme ratio. Problems in mixing and mass transfer are other limitations associated with high substrate concentration [18]. The sugar production and saccharification rate were found to increase with increase in the concentration of alkaline-treated date palm leaves with an optimum of 4% [19]. Another study on enzymatic hydrolysis of alkaline-pretreated wheat straw, rice straw and bagasse by *Bacillus subtilis* reported an optimum substrate concentration of 4% for best results [36]. Maximum hydrolysis of alkaline-treated sugarcane bagasse was reached at an optimum substrate concentration of 2.5% [50]. Krishna and Chowdary [51] and Wen et al. [52] reported that an increase in substrate concentration decreased the rate of saccharification, probably as a result of inhibition of enzymes by the products of saccharification or reduced synergistic action between enzymes. Furthermore, Zhang et al. [53] reported that maximum saccharification of steam explosion-pretreated corn stover was obtained at 3% substrate concentration and a further increase in substrate concentration decreased the rate of hydrolysis.

6.5.2 Enzyme Concentration

The enzyme-to-substrate ratio used during saccharification process is another significant factor affecting the rate of enzymatic hydrolysis. A high enzyme dosage considerably improves the sugar yield but would also significantly raise procedure costs [12]. Additionally, high cellulase concentrations may slow down the saccharification process due to increase in the rate of transglycosylation reactions, improper mixing and suspension of slurry [36]. The enzyme dosage varies widely depending on the purpose of the enzymatic saccharification and is usually in the range 5–35 FPU per gram substrate. Cellulase concentration of 120 FPU/g substrate was found to be optimal for enzymatic saccharification, and on increasing the concentration of enzyme from 40 to 120 FPU, only a 10% increase in the sugar yield was observed [51]. Zhang et al. [53] reported the optimum enzyme concentrations for saccharification of ammonia-pretreated corncob and steam explosion-pretreated corn stover as 50 and 20 FPU/g substrate, respectively. Contrary to these reports, a study reported that the enzyme dosage for saccharification reaction should not exceed 5–8 FPU/g substrate and reported the maximum saccharification yield (90%) from corncob residue at the enzyme concentration of 8 FPU/g substrate [54]. A study on the interaction of cellulase, xylanase and β-glucosidase on enzymatic hydrolysis of alkali-treated water hyacinth was carried out and the maximum yield of total reducing sugar with a value of 431.421 mg/g was obtained at a cellulase concentration of 15 U/g, xylanase concentration of 300 U/g and β-glucosidase concentration of 59.68 U/g [55]. The effect of enzyme concentration (10–80 FPU/g substrate) on the enzymatic hydrolysis of date palm wastes was investigated and the optimum concentration was found to be 30 FPU/g substrate with maximum saccharification percentage and sugar production as 71.03% and 31.57 mg/mL, respectively [19]. The results revealed that the effectiveness of sugar production and saccharification rate depended strongly on the enzyme–substrate ratio. Enzyme loadings are described as activity assays units such as filter paper units (FPUs) and CMCase (carboxymethyl cellulase) activity, mg enzyme protein per g of glucan or biomass and simple gravimetric or volumetric loadings. Most of the commercial enzyme preparations are provided in the form of liquid and the solvent and additives present in them must be taken into account while using gravimetric and volumetric loadings. The enzyme content in commercial enzymes like Celluclast 1.5 L and Novozym 188 is reported to be about 4.0 and 5.1% (w/v), respectively [15].

As the cost of cellulase contributes significantly to the cost of the hydrolysis process, two aspects are globally addressed to reduce the cost of enzymes during the production of fuel ethanol from lignocellulosic feedstocks: optimization of various factors affecting the process of cellulases production and development of a more efficient cellulase-based catalysis system. Additionally, development of more efficient thermophilic cellulases, recovery and recycling of the enzymes are other attractive strategies to cut down enzymatic hydrolysis costs [18, 56]. The recycling of enzymes is largely influenced by enzyme adsorption onto the solid residuals, especially to lignin, enzymes dissolution in the hydrolyzates and enzymes inactivation. Cellulase enzymes can be recovered using immobilization technique based on the principle of fixation of the enzymes onto a solid matrix either by adsorption or grafting and thus enables the separation of enzymes from process flow [56]. Co-immobilization of enzyme with the fermenting micro-organism to improve the overall cellulose to ethanol bioconversion is also an option. Major limitation of immobilization technique is the separation of immobilized matrix from the residual solid. However, it can be overcome by using magnetic particles like magnetic chitosan microspheres [57]. Another strategy for recycling of these enzymes involves concentration of the filtrate obtained after hydrolysis of the cellulose fraction of the biomass using ultrafiltration process to remove sugars and other small inhibitory compounds that may hinder the enzymes' action during further hydrolysis [58]. The readsorption process of cellulase during enzymatic hydrolysis of corncob was investigated and after 96 h of ethanol fermentation, 62.5% of the 40% of the initial cellulase that remained in the broth was recycled and reused in fresh substrate readsorption for 90 min. Twofold increase in ethanol output was achieved after two cycles of fermentation and readsorption under optimum conditions of pH 5.0, 155 dry biomass loading and 45 FPU/g glucan of cellulase loading [59].

6.5.3 Pretreatment Method

The efficiency of enzymatic hydrolysis of cellulosic substrates depends on the structural complexity lignocellulosic substrate, including lignin content, cellulose crystallinity, degree of polymerization and surface area [18, 57]. Lignin acts as a shield and interferes with hydrolysis by preventing access of cellulases and hemicellulases to cellulose and hemicellulose fraction of the biomass, thereby resulting in extended reaction times to achieve high conversion rate. Another obstacle in enzymatic hydrolysis process is irreversible adsorption of a large amount of the cellulase enzyme by lignin rendering it unavailable for further hydrolysis process. Therefore, degradation and removal of lignin is one of the most significant steps for conversion of cellulose and ethanol and is achieved by various pretreatment methods to enhance the process of enzymatic hydrolysis [12]. Additionally, removal of hemicellulose is also required to increase the mean pore size of the substrate to further increase the accessibility of cellulase enzyme to cellulose.

A study on use of pretreatment methods to enhance saccharification process reported that pretreatment with 2% NaOH was found to be the most effective method prior to the enzymatic saccharification of wheat straw, rice straw and bagasse by *Bacillus subtilis* cellulases, obtaining saccharification rates of 33.0%, 25.5% and 35.5%, respectively [36]. These results are in concord with those of another report stating that alkaline pretreatment of poppy stalks resulted in an increased saccharification degree [60]. Enzymatic saccharification of untreated, acid-treated and alkali-treated water hyacinth biomass was performed using cellulase, xylanase and β-glucosidase and it was found that alkaline pretreatment prior to enzymatic treatment resulted in highest yield of both reducing sugar and xylose [55]. Enzymatic hydrolysis of water hyacinth (*Eichhornia crassipes*) and water lettuce (*Pistia stratiotes* L.) was performed and twenty pretreatment methods were used and examined for their effect on the efficiency of enzymatic hydrolysis. The alkaline/oxidative (A/O) pretreatment method using sodium hydroxide and hydrogen peroxide was found to be the most effective in terms of reducing sugars production [61]. Pretreatment of *Eichhornia crassipes* and wheat straw using crude glycerol, ionic liquids pretreatment and conventional dilute acid pretreatment was evaluated and compared for their effect on enzymatic hydrolysis yield of pretreated samples. Ionic liquid pretreatment using 1-butyl-3methylimidazolium acetate (BMIMA) was found to be most effective in improving the saccharification yield of wheat straw and produced double glucose and more than triple total reducing sugar yields compared to conventional dilute acid pretreatment method. In case of water hyacinth, no significant difference in the amount of both glucose and reducing sugar was observed on pretreatment with crude glycerol, pure glycerol and dilute acid. However, crude glycerol was preferred due to its cost-effectiveness, easy availability being a waste of diesel industry and no requirement of alkali for neutralization after pretreatment [62].

The effect of different pretreatments on enzymatic saccharification of date palm wastes for its subsequent exploitation as a substrate for bioethanol production was investigated. The results indicated that the maximum saccharification degrees and glucose production for untreated substrates were much lower compared to the treated biomass. The low saccharification rate in untreated substrate may be attributed to the high lignin content, which hinders the large molecules of enzymes from penetrating into the complex network of carbohydrates that constitute the plant cell wall. Among the various pretreatment methods employed, the maximum saccharification rate (41.40%) and glucose production (9.2 mg/mL) were obtained with alkaline-treated date palm leaves. Acid pretreatment method was found to be the next best alternative with maximum saccharification degree (19.57%) and glucose production (4.37 mg/mL) with treated date palm [19]. Total reducing sugar concentration in the enzymatic hydrolysates of bagasse was determined after various alkaline and ultrasonic pretreatment conditions, and the highest yield of 36.21 ± 0.46% was obtained on pretreatment with 1% NaOH followed by ultrasonic treatment for 60 min [63]. On the contrary, another study reported the maximum saccharification rate for enzymatic hydrolysis of cellulosic wastes when pretreated by diluted H_2SO_4 at an elevated temperature and was approximately 80% [64]. These reports indicated that no universal pretreatment can be used for all the lignocellulosic substrates and parameters like polymer composition of biomass, degree of polymerization and crystallinity of cellulose, lignin composition, ferulate and coumarate cross-linking and side-chain branching of hemicelluloses should be considered for the optimization and selection of best pretreatment method.

6.5.4 Temperature

The temperature of the hydrolysis reaction is one of critical factors exerting profound influence on the end product and is a function of the substrate; the enzyme source and hydrolysis reaction time. Increase of temperature beyond an optimal range may result in thermal denaturation and the loss of biological activity of the enzyme. The effect of temperature is usually dependent on the strain of micro-organism used for cellulase production and enzymatic hydrolysis. Maximum CMCase activity in *Cellulomonas*, *Bacillus* and *Micrococcus* sp. was obtained at a reaction temperature of 40 °C [65]. Enzymatic hydrolysis of alkali-pretreated sugarcane bagasse using crude enzyme extract from *Trichoderma* sp. achieved highest rate of hydrolysis (37.29%) at 50 °C [66]. Ahmed et al. [67] reported that the highest sugar yield of about 343 mg/g dry substrate was produced from NaOH-pretreated wheat straw using cellulase from *Penicillium waksmanii* at 55 °C after 30 h. The rate of enzymatic hydrolysis of cellulose increased gradually till 50 °C and then dropped drastically due to the loss in enzyme activity at higher temperature [68]. These results are consistent with other reports wherein maximum enzyme adsorption and saccharification rate were achieved at 50 °C [36, 50, 51]. The optimal temperature for enzymatic hydrolysis of steam explosion-pretreated leaves and pseudostem of banana and corn stover was found to be 45 °C and 40 °C, respectively [53, 69]. The reducing sugars formation on enzymatic hydrolysis of the filter paper, eucalyptus sawdust and corncob using cellulolytic complex from *Trichoderma viride* was maximum at a temperature of 40 °C [21]. Evaluation of optimal temperature for enzymatic saccharification of alkali-treated date palm leaves was performed at 40–70 °C and the maximum enzymatic hydrolysis was attained at 50 °C, corresponding to a saccharification degree of 71.23%. Further increase in hydrolysis temperature decreased the saccharification rate to 66.48% at 55 °C and 27.31% at 70 °C as a result of an increase in enzyme deactivation at higher temperatures [19].

6.5.5 pH

Initial pH of the fermentation medium is one of the most significant factors which influence enzymatic hydrolysis by causing conformational change in the active site of an enzyme and thus altering the 3-D structure of enzymes. The effect of the pH on the enzymatic hydrolysis is due to variations in the ionization state of the components of the cellulolytic system with the pH variation. Most microbial extracellular enzymes are produced at an optimum growth pH and extremely high and low pH values outside the optimal range result in a complete loss of activity for most of the enzymes. Cellulase production, enzymatic processes and transport of compounds across the cell membrane are markedly dependent on the hydrogen ion concentration of the production medium. Maximum cellulase production for several bacterial species usually occur in pH range of 7–9 as depicted in various reports on cellulase production and enzymatic hydrolysis by *Bacillus* sp., *Clostridium acetobutylium* and *straminisolvens* sp. [70–72]. Optimum pH for fungal growth and enzyme biosynthesis varies in different species and is usually in slightly acidic range. Also, Krishna and Chowdary [51] reported that the initial pH has a profound effect on both adsorption and hydrolytic capacity of cellulases by influencing the formation of enzyme–substrate complex, a prerequisite for the hydrolysis reaction that usually occur at around pH 4.8. Maximum cellulase production from *Trichoderma reesei* RUT-C30 was reported at pH 4.8 while for *Trichoderma lignorum*, it was maximum at pH 6.0 [69, 73]. The optimum pH for cellulase activity of *Aspergillus niger* was found in the range of pH 4–7 as reported in several studies [74, 75]. Maximum enzymatic hydrolysis rate of sugarcane bagasse after alkali pretreatment was reported at pH 5.0 [50]. The reducing sugars formation on enzymatic hydrolysis of eucalyptus sawdust, and corncob using cellulolytic complex from *Trichoderma viride* was maximum at pH 5.2 [21]. The sugar and saccharification percentages after enzymatic hydrolysis of date palm leaves using the cellulases obtained from *Geobacillus stearothermophilus* were highest at pH 5.0 [19]. Contrary to popular belief, enzymatic saccharification of various agro-residues such as orange peel, water hyacinth, sugarcane bagasse, dried leaves and flowers, and coconut shell from *Rhizopus oryzae* PR 7 was optimized to be best at pH 7.0, higher than the acidic pH optima of CMCase enzyme produced by *Trichoderma viride* [76]. The pH of the fermentation medium tends to increase in the beginning due to utilization of hemicellulose and amorphous cellulose content of the lignocellulosic biomass for fungal growth. However, in later stages, the formation of carboxylic groups and carbonic acids from lignin contributes to a drop in the pH of the culture medium [77].

6.5.6 Reaction Time

The reaction/incubation time for the hydrolysis reaction is of critical significance and is usually dependent on the complexity of substrate; the enzyme source and pH of fermentation medium. Krishna and Chowdary [51] reported that the optimum reaction time for enzymatic hydrolysis of alkaline-pretreated *Antigonun leptopus* leaves was 24 h and further incubation did not show any significant effect on hydrolysis rate or sugar production. Enzymatic hydrolysis of alkali-treated sugarcane bagasse increased at a higher pace in the first 8 h with a significant decrease in the rate at the later stages of process [78]. The optimum reaction period for enhanced sugar production and saccharification rate of alkali-treated date palm was determined and found to be at maximum value after 24 h with no further effect on saccharification on extending the reaction time. Enzymatic hydrolysis of wheat straw, rice straw and bagasse by *Bacillus subtilis* showed maximum saccharification degree after 20 h [36]. Also, another study reported that though the saccharification rate increased in a uniform pattern up to 24 h, the saccharification per unit of time decreased probably because of cellulose recrystallization and selective enzyme attack on the amorphous regions of cellulose [79]. The enzymatic hydrolysis of water hyacinth hydrolysate was conducted using commercial cellulase (ACCELLERASE1500) at a loading of 0.30 ml/g water hyacinth and the highest reducing sugar concentration of 11.95 g/L was produced at 50 °C after 48 h [80]. Sartori et al. [21] verified that the reducing sugars formation on enzymatic hydrolysis of the filter paper, eucalyptus sawdust and corncob showed gradual increase until 12 h of saccharification remaining stable after this period with both cellulolytic complex obtained from *Trichoderma viride* and commercial cellulase. The best saccharification yields on hydrolysis of corncob residue and corn stover were reported at 48 h of reaction time [53, 54].

References

1. Gandla ML, Martin C, Jonsson LJ (2018) Analytical enzymatic saccharification of lignocellulosic biomass for conversion to biofuels and bio-based chemicals. Energies 11:2936
2. Ogeda TL, Petri DFS (2010) Hidrolise Enzimatica de Biomassa. Qumica Nova 33(7):1549–1558

3. Van DJS, Pletschke BI (2012) A review of lignocellulose bioconversion using enzymatic hydrolysis and synergistic cooperation between enzymes-factors affecting enzymes, conversion and synergy. Biotechnol Adv 30:1458–1480
4. Kubicek CP (2013) Systems biological approaches towards understanding cellulase production by *Trichoderma reesei*. J Biotechnol 163:133–142
5. Duff SJB, Murray WD (1996) Bioconversion of forest products industry waste cellulosics to fuel ethanol-a review. Bioresour Technol 55:1–33
6. Himmel ME, Ruth MF, Wyman CE (1999) Cellulase for commodity product from cellulosic biomass. Curr Opin Biotechnol 10:358–364
7. Horn SJ, Vaaje-Kolstad G, Westereng B, Eijsink VGH (2012) Novel enzymes for the degradation of cellulose. Biotechnol Biofuels 5:45
8. Morgenstern I, Powlowski J, Tsang A (2014) Fungal cellulose degradation by oxidative enzymes: From dysfunctional GH61 family to powerful lytic polysaccharide monooxygenase family. Brief Funct Genom 13:471–481
9. Kuhad RC, Mehta G, Gupta R, Sharma KK (2011) Fed batch enzymatic saccharification of newspaper cellulosics improves the sugar content in the hydrolysate and eventually the ethanol fermentation by *Saccharomyces cerevisiae*. Biomass Bioenergy 34:1189–1194
10. Kuhad RC, Gupta R, Khasa YP, Singh A (2011) Bioconversion of pentose sugars to ethanol: current and future prospects. Renew Sustain Energy Rev 15:4950–4962
11. Ooshima H, Sakata M, Harano Y (1986) Enhancement of enzymatic hydrolysis of cellulose by surfactant. Biotechnol Bioeng 28:1727–1734
12. Prasad S, Singh A, Joshi HC (2007) Ethanol as an alternative fuel from agricultural, industrial and urban residues. Resour Conserv Recycl 50:1–39
13. Sanchez OJ, Cardona CA (2008) Trends in biotechnological production of fuel ethanol from different feedstocks. Bioresour Technol 99:5270–5295
14. Xu H, Jia S, Liu J (2011) Development of a mutant strain of *Bacillus subtilis* showing enhanced production of acetoin. African J Biotechnol 10:779–788
15. Pareek N, Gillgren T, Jonsson LJ (2013) Adsorption of proteins involved in hydrolysis of lignocellulose on lignins and hemicelluloses. Bioresour Technol 148:70–77
16. Bhunia B, Behera KK, Ahmed AB, Sharma HP (2010) Optimization of alkaline protease activity from bacillus subtilis 2724 by response surface methodology (RSM). Int J Biolog Sci Eng 1(3):158
17. Bisaria VS (1991) Bioprocessing of agro-residues to glucose and chemicals, bioconversion of waste materials to industrial products. Elsevier, London, pp 210–213
18. Sun Y, Cheng J (2002) Hydrolysis of lignocellulosic materials for ethanol production: a review. Bioresour Technol 83(1):1–11
19. Alrumman SA (2016) Enzymatic saccharification and fermentation of cellulosic date palm wastes to glucose and lactic acid. Brazilian J Microbiol 47:110–119
20. Sukumaran RK, Singhania RR, Mathew GM, Pandey A (2009) Cellulase production using biomass feed stock and its application in lignocellulose saccharification for bio-ethanol production. Renew Energy 34(2):421–424
21. Sartori T, Tibolla H, Prigol E, Colla LM, Costa JAV, Bertolin TE (2015) Enzymatic saccharification of lignocellulosic residues by cellulases obtained from solid state fermentation using *Trichoderma viride*. BioMed Res Int Article ID 342716
22. Kuhad RC, Manchanda M, Singh A (1999) Hydrolytic potential of cellulolytic enzymes from a mutant strain of *Fusarium oxysporum*. Bioproc Eng 20:133–135
23. Pandey A, Radhakrishnan S (1993) The production of glucoamylase by *Aspergillus niger* NCIM 1245. Proc Biochem 28:305
24. Mudgett RE (1986) Solid state fermentation. In: Demain AL, Solomon NA (eds) Manual of industrial microbiology and biotechnology. Washington DC, Am Soc Microbiol, pp 66–83
25. Castilho LR, Polato CMS, Baruque EA, Sant Anna GL, Jr Freire DMG (2000) Economic analysis of lipase production by *Penicillium restrictum* in solid-state and submerged fermentations. Biochem Eng J 4:239–247

26. Aguilar CN, Rodriguez R, Gutierrez-Sanchez G, Augur C, Favela-Torres E, Prado-Barragan LA, Ramırez-Coronel A, Contreras-Esquivel JC (2007) Microbial tannases: advances and perspectives. Appl Microbiol Biotechnol 76(1):47–59
27. Pandey A, Soccol CR, Mitchell D (2000) New developments in solid state fermentation. I-bioprocesses and products. Process Biochem 35(10):1153–1169
28. Couto SR, Sanroman MA (2006) Application of solid-state fermentation to food industry—a review. J Food Eng 76(3):291–302
29. Krishna C (2005) Solid-state fermentation systems-an overview. Crit Rev Biotechnol 25:1–30
30. Berlin A, Maximenko V, Gikes N, Saddle J (2007) Optimization of enzyme complexes for lignocellulose hydrolysis. Biotechnol Bioeng 97(2):287–296
31. Kuo CH, Lee CK (2009) Enhancement of enzymatic saccharification of cellulose by cellulose dissolution pretreatments. Carbohyd Polym 77(1):41–46
32. Santos ALF, Kawase KYF, Coelho GLV (2010) Enzymatic saccharification of lignocellulosic materials after treatment with supercritical carbon dioxide. J Supercrit Fluids 56:277–282
33. Singh A, Bishnoi NR (2013) Comparative study of various pretreatment techniques for ethanol production from water hyacinth. Ind Crop Prod 44:283–289
34. Annamalai N, Rajeswari MV, Balasubramanian T (2014) Enzymatic saccharification of pretreated rice straw by cellulase produced from *Bacillus carboniphilus* CAS 3 utilizing lignocellulosic wastes through statistical optimization. Biomass Bioenergy 68:151–160
35. Dadi AP, Varanasi S, Schall CA (2006) Enhancement of cellulose saccharification kinetics using an ionic liquid pretreatment step. Biotechnol Bioeng 95(5):904–910
36. Akhtar MS, Saleem M, Akhtar MW (2001) Saccharification of lignocellulosic materials by the cellulases of *Bacillus subtilis*. Int J Agric Biol 3:199–202
37. Berlin A, Maximenko V, Bura R, Kang KY, Gilkes N, Saddler J (2006) A rapid microassay to evaluate enzymatic hydrolysis of lignocellulosic substrates. Biotechnol Bioeng 93:880–886
38. Normark M, Pommer L, Grasvik J, Hedenstrom M, Gorzsas A, Winestrand S, Jonsson LJ (2016) Biochemical conversion of torrefied Norway spruce after pretreatment with acid or ionic liquid. Bioenergy Res 9:355–368
39. Gomez LD, Whitehead C, Roberts P, McQueen-Mason SJ (2011) High-throughput saccharification assay for lignocellulosic materials. J Vis Exp 53:1–5
40. Wang Z, Wu G, Jonsson LJ (2018) Effects of impregnation of softwood with sulfuric acid and sulfur dioxide on chemical and physical characteristics, enzymatic digestibility, and fermentability. Bioresour Technol 247:200–208
41. Normark M, Winestrand S, Lestander TA, Jonsson LJ (2014) Analysis, pretreatment and enzymatic saccharification of different fractions of Scots pine. BMC Biotechnol 14:20
42. Maurya DP, Singh D, Pratap D et al (2012) Optimization of solid state fermentation conditions for the production of cellulase by *Trichoderma reesei*. J Environ Biol 33:5–8
43. Amira D, Roshanida AR, Rosli MI (2012) Effects of xylanase and cellulase production during composting of EFB and POME using fungi. Intl J Biol Biomol Agric Food Biotechnol Eng 6:340–343
44. Kumar D, Ashfaque M, Muthukumar M et al (2012) Production and characterization of carboxymethyl cellulase from *Paenibacillus polymyxa* using mango peel as substrate. J Environ Biol 33:81–84
45. Heck JX, Hertz PF, Ayub MAZ (2002) Cellulase and xylanase production by isolated amazon *Bacillus* strains using soybean industrial residue based solid-state cultivation. Brazilian J Microbiol 33:213–218
46. Poorna AA, Prema P (2007) Production of cellulase-free endoxylanase from novel alkalophillic thermostable *Bacillus pumilis* by solid-state fermentation and its application in wastepaper recycling. Bioresour Technol 98:485–490
47. Kang SW, Park YS, Lee JS, Hong SI, Kim SW (2004) Production of cellulases and hemicellulases by *Aspergillus niger* KK2 from lignocellulosic biomass. Bioresour Technol 91:153–156
48. Swelim M, Hammad AI, Gannam RB et al (2010) Some critical factors affecting cellulase (S) production by *Aspergillus terreus* Mam-F23 and *Aspergillus flavus* Mam-F35 under solid-state fermentation of wheat straw. World Appl Sci J 9:1171–1179

49. Mahalakshmi N, Jayalakshmi S (2016) Cellulase production by *Aspergillus niger* under solid state fermentation using agro industrial wastes. Int J Adv Multidiscip Res 3:78–83
50. Mahamud MR, Gomes DJ (2012) Enzymatic saccharification of sugar cane bagasse by the crude enzyme from indigenous fungi. J Sci Res 4:227–238
51. Hari Krishna S, Chowdary GV (2000) Optimization of simultaneous saccharification and fermentation for the production of ethanol from lignocellulosic biomass. J Agric Food Chem 48:1971–1976
52. Wen Z, Liao W, Chen S (2004) Hydrolysis of animal manurelignocellulosics for reducing sugar production. BioresourTechnol 91:31–39
53. Zhang Y, Lu C, Tang J et al (2011) Enhanced saccharification of steam explosion pretreated corn stover by the supplementation of thermo acidophilic β-glucosidase from a newly isolated strain, *Tolypocladium cylindrosporum* syzx4. Afr J Microbiol Res 5:2413–2421
54. Ouyang J, Li Z, Li X, Ying H, Yong Q (2009) Enhanced enzymatic conversion and glucose production via two-step enzymatic hydrolysis of corncob residue from xylo-oligo saccharides producer's waste. Biol Res 4:1586–1599
55. Ganguly A, Halder S, Laha A, Saha N, Chatterjee PK, Dey A (2013) Effect of alkali pretreatment on water hyacinth biomass for production of ethanol. Adv Chem Eng Res 2(2):40
56. Mosier NS, Wyman C, Dale B, Elander R, Lee YY, Holtzapple M, Ladisch MR (2005) Features of promising technologies for pretreatment of lignocellulosic biomass. Bioresour Technol 96:673–686
57. Taherzadeh MJ, Karimi K (2008) Pretreatment of lignocellulosic wastes to improve ethanol and biogas production: a review. Int J Mol Sci 9:1621–1651
58. Tu MB, Chandra RP, Saddler JN (2007) Evaluating the distribution of cellulases and the recycling of free cellulases during the hydrolysis of lignocellulosic substrates. Biotechnol Prog 23:398–406
59. Du R, Su R, Zhang M, Qi W, He Z (2014) Cellulase recycling after high-solids simultaneous saccharification and fermentation of combined pretreated corncob. Fron Energy Res. https://doi.org/10.3389/fenrg.2014.00024
60. Uzunlu N, Hos gün EZ, Bozan B (2014) Optimization of alkaline pretreatment for enzymatic saccharification of poppy stalks. BioResources 9:2824–2834
61. Mishima D, Tateda M, Ike M, Fujita M (2006) Comparative study on chemical pretreatments to accelerate enzymatic hydrolysis of aquatic macrophyte biomass used in water purification processes. Bioresour Technol 97:2166–2172
62. Guragain YN, Coninck JD, Husson F, Durand A, Rakshit SK (2011) Comparison of some new pretreatment methods for second generation bioethanol production from wheat straw and water hyacinth. Bioresour Technol 102:4416–4424
63. Soontornchaiboon W, Kim SM, Pawongrat R (2016) Effects of alkaline combined with ultrasonic pretreatment and enzymatic hydrolysis of agricultural wastes for high reducing sugar production. Sains Malaysiana 45(6):955–962
64. Saha BC, Iten LB, Cotta MA, Wu YV (2005) Dilute acid pretreatment, enzymatic saccharification and fermentation of rice hulls to ethanol. Biotechnol Prog 21:816–822
65. Immanuel G, Dhanusa R, Prema P, Palavesam A (2006) Effect of different growth parameters on endoglucanase enzyme activity by bacteria isolated from coir retting effluents of estuarine environment. Int J Environ Sci Technol 3:25–34
66. Wang W, Kang L, Wei H et al (2011) Study on the decreased sugar yield in enzymatic hydrolysis of cellulosic substrate at high solid loading. Appl Biochem Biotechnol 164:1139–1149
67. Ahmed FM, Rahman SR, Gomes DJ (2012) Saccharification of sugarcane bagasse by enzymatic treatment for bioethanol production. Malays J Microbiol 8:97–103
68. Gokhale AA, Lu J, Lee I (2013) Immobilization of cellulase on magneto responsive graphene nano-supports. J Mol Catal B Enzym 90:76–86
69. Baig MMV, Baia MLB, Baig MIA, Yasmeen M (2004) Saccharification of banana agro-waste by cellulolytic enzymes. Afr JBiotechnol 3:447–450
70. Song FL, Forsberg CW, Gibbins LN (1985) Cellulolytic activity of *Clostridium acetobutylium*. Appl Environ Microbiol 50:220–228

71. Yang VW, Zhuang Z, Elegir G, Jeffries TW (1995) Alkaline-active xylanase produced by an alkaliphilic *Bacillus* sp isolated from kraft pulp. J Ind Microbiol 15:434–441
72. Souichiro K, Shin H, Zong JC, Masaharu I, Akira Y, Yasuo I (2004) *Clostridium straminisolvens* sp. nov., a moderately thermophilic, aerotolerant and cellulolytic bacterium isolated from a cellulose-degrading bacterial community. Int J Sys Evol Microbiol 54:2043–2047
73. Das M, Banerjee R, Bal S (2008) Multivariable parameter optimization for the endoglucanase production by *Trichoderma reesei* Rut C30 from *Ocimum gratissimum* seed. Brazilian Arch Biol Technol 51:35–41
74. Akiba S, Kimura Y, Yamamoto K, Kumagai H (1995) Purification and characterization of a protease-resistant cellulase from *Aspergillus niger*. J Ferment Bioeng 79:125–130
75. Sohail M, Siddiqi R, Ahmad A, Khan SA (2009) Cellulase production from *Aspergillus niger* MS82: effect of temperature and pH. New Biotechnol 25:437–441
76. Karmakar M, Ray R (2011) Current trends in research and application of microbial cellulases. Res J Microbiol 6:41–53
77. Mojović L, Nikolić S, Rakin M et al (2006) Production of bioethanol from corn meal hydrolyzates. Fuel 85:1750–1755
78. Kim DW, Yang JH, Jeong YK (1988) Adsorption of cellulase from *Trichoderma viride* on microcrystalline cellulose. Appl Microb Biotechnol 28:148–154
79. Portjanskaja E, Preis S, Kallas J (2006) Aqueous photocatalytic oxidation of lignin and humic acids with supported TiO_2. Int J Photoene 8:1–7
80. Jongmeesuk A, Sanguanchaipaiwong V, Ochaikul D (2014) Pretreatment and enzymatic hydrolysis from water hyacinth (*Eichhornia crassipes*). KMITL Sci Technol J 14(2):79–86

Chapter 7
Bioethanol Production from Water Hyacinth

Abstract Enhanced development of bioethanol for its use as a carbon-neutral renewable and clean fuel is ever-increasing as it reduces CO_2 emissions and associated climate change; it is used as fuel mix and octane enhancer in gasoline and improves the ambient air quality. In comparison with conventional agricultural biomass, lignocellulose biomass is a potential and sustainable substrate for bioethanol production. Lignocellulose biomass commonly comes from terrestrial origins, but aquatic plants and weeds like water hyacinth can also provide adequate biomass for bioethanol production. Aquatic feedstock like water hyacinth has a decentralized availability and thus requires a distributed production strategy for efficient availability. Water hyacinth has been widely exploited for biofuel production with research dedicated to efficient pretreatment, enzymatic hydrolysis and fermentation strategies. Novel processes like CBP can increase the bioethanol conversion yield to about 92% of the theoretical value compared to ~50% by conventional methods along with a reduction in operational costs and inhibitors. Additionally, recent advances in genetically engineering have shown promising results for the generation of fermenting organisms with higher alcohol productivity and tolerance. The current chapter discusses bioethanol production from water hyacinth, various fermentation strategies and their limitations, microbes in fermentation and current advances and commercial status of bioethanol production.

Keywords Bioethanol · Biomass · Fermentation · Lignocellulose · Saccharification · Yeasts

Bioethanol production from biomass is a promising alternative new energy source that has a high potential to substitute fossil fuels. In 2010, 12.7 billion tons of oil equivalents [1] were consumed globally which included 32.4% oil, 27.3% coal and peat, 21.4% natural gas, and 10% biofuels with transport sector dominating with 61.5% of the total oil consumption. Therefore, renewable alternatives for the transport sector are to be explored seriously for the replacement of fossil fuels. Bioethanol is chiefly produced from starch of corn, sugar beet, and sugarcane and lignocellulosic biomass. Second-generation bioethanol from lignocellulosic sources offers the opportunity to reduce the carbon dioxide emissions, mitigate air pollution, dependence on energy imports, and diversification of energy matrix. Therefore, complex

A. Sharma and N. K. Aggarwal, *Water Hyacinth: A Potential Lignocellulosic Biomass for Bioethanol*, https://doi.org/10.1007/978-3-030-35632-3_7

lignocellulosic biomass has been presented as a feasible alternative due to its major advantages like abundant availability in nature and large quantities generated in the form of agricultural and industrial waste, sustainability, recyclability, carbon neutrality, higher cellulose and hemicellulose content and compositional uniformity. Conversion of residual biomass into other value-added chemicals with minimum waste generation as a part of integrated biorefinery facility is another advantage of using lignocellulosic substrates.

7.1 Fermentation Strategies

Ethanol fermentation is a biological process during which simple sugars are acted upon and utilized by microorganisms to convert them into ethanol and CO_2. The fermentation of lignocellulosic hydrolysate is a complex process and reducing sugars produced after enzymatic hydrolysis can be further fermented to ethanol by microorganisms using a variety of approaches. Fermentation systems for conversion of lignocelluloses to ethanol depend on the nature of the feedstock and microorganism employed and include (a) separate hydrolysis and fermentation (SHF), (b) simultaneous saccharification and fermentation (SiSF), (c) simultaneous saccharification and co-fermentation (SSCF), (d) consolidated bioprocessing (CBP).

7.1.1 Separate Hydrolysis and Fermentation (SHF)

In SHF, enzymatic hydrolysis of lignocellulosic materials is performed separately from fermentation. The pretreated biomass first undergoes the process of saccharification followed by fermentation to ethanol [2]. A major advantage of SHF is that the separation of enzymatic hydrolysis and fermentation process which allows enzymes during saccharification to be operated at high temperature for better performance while fermenting organisms can be provided optimum moderate temperature for efficient sugar utilization. Also, this technique facilitates the selection of sugar-appropriate microorganisms to ferment the different sugars. End-product inhibition of enzymes by cellobiose and glucose, microbial contamination and higher investment costs for two separate reactors are however, limitations of this strategy [3, 4]. Cellobiose reduces cellulase activity by 60% at a concentration of as low as 6 g/L while glucose reduces β-glucosidase activity by 75% at a concentration of 3 g/L. Enzymatic hydrolysis of sugarcane bagasse, rice straw and water hyacinth biomass was carried out using 50 FPU cellulase and 10 U of β-glucosidase produced using *Trichoderma reesei* RUT C30 and *Aspergillus niger* MTCC 7956, respectively, and enzymatic hydrolysate of rice straw containing highest yield of reducing sugars (26.3 g/L) was used as substrate for ethanol production by *Saccharomyces cerevisiae* producing 0.093 g ethanol per gram of pretreated rice straw [5]. The production of bioethanol of enzymatic hydrolysate of steam-exploded material pine wood chip produced after saccharification using Cellic™ CTec2 was performed by *Saccharomyces cerevisiae* Red Ethanol® and resulted in 10.6 g of ethanol per 100 g of a dry substrate [6].

7.1.2 Simultaneous Saccharification and Fermentation (SiSF)

In SiSF, enzymatic hydrolysis and fermentation are consolidated in a single process so that the reducing sugars produced are simultaneously consumed by the fermenting organism for their conversion to ethanol. SiSF was first demonstrated by Gauss and co-workers in 1976 [7]. The fermentation of microwave/lightwave-assisted alkali pretreated water hyacinth was studied in the simultaneous saccharification and fermentation mode (SSF). The ethanol yield reached 0.132 g/g substrate and 0.137 g/g substrate with *Saccharomyces cerevisiae* and *Pichia stipitis*, respectively [8]. Various advantages of SiSF over SHF process include a lower enzyme requirement, neutralization of end-product inhibition, reduced risk of contamination, shorter process time, improved ethanol yield and productivity, and cost-effectiveness [2–4]. However, limited yeast recirculation due to the presence of lignin residues and cellulase inhibition by ethanol is major limitations [9]. Another obstacle associated with SiSF is the difference in optimum temperature condition for enzyme hydrolysis (45–50 °C) and fermentation organism (30–35 °C) making hydrolysis a rate-limiting process [3, 10]. Use of thermotolerant organisms like *Candida acidothermophilum* and *Kluyveromyces marxianus* to overcome temperature limitation has also been proposed [11].

7.1.3 Simultaneous Saccharification and Co-fermentation (SSCF)

Simultaneous saccharification and co-fermentation (SSCF) as an upgrade of SiSF involves two process steps: cellulase production in the first step and enzymatic hydrolysis and fermentation of both hexose and pentose sugars in a single bioreactor in the second step [12]. SSCF is a process that offers a better opportunity for more streamlined processing due to the simultaneous enzymatic hydrolysis and fermentation process. Also, co-fermentation of pentoses and hexoses by a single strain of microorganisms significantly improves the cost-effectiveness of the process for commercial production of lignocellulosic bioethanol. A recombinant strain of *Zymomonas mobilis* capable of fermenting both C5 and C6 sugars has been used to produce ethanol from a variety of substrates like hardwood, yellow poplar, hybrid poplar wood and sugarcane bagasse [11]. Ethanol production from corn stover using a recombinant strain of *E. coli* was evaluated, and ethanol yield of 109% of theoretical yield from glucose fermentation was obtained [13]. The success of SSCF is generally limited by the lack of microorganisms that can co-ferment glucose and xylose concurrently. Advantages of SSCF over SHF include low cost, higher ethanol yield and shorter processing time.

7.1.4 Consolidated Bioprocessing (CBP)

Consolidated bioprocessing (CBP) involves a single process step of cellulolytic enzymes complex production, hydrolysis and fermentation of both hexose and pentose sugars by a single type of microorganisms in a single bioreactor. The process is also known as direct microbial conversion (DMC). CBP implies no operating costs or capital investment for enzymes purchase or production and a reduction in substrate utilization for enzymes production [14, 15]. However, determining compatible combinations of microorganisms which ferment cellulose to ethanol is a major limitation and requires extensive research. Ethanol production by CBP has been explored by thermophilic bacteria like *Clostridium thermocellum* and yeasts like *Neurospora crassa*, *Monilia* sp., *Paecilomyces* sp. but was found to produce a low ethanol yield along with the formation of inhibitory by-products [16].

7.2 Microbes in Fermentation

Microbial utilization of sugars for ethanol production has been explored extensively by researchers [4, 17, 18]. However, unlike sucrose- and starch-based bioethanol produced from one or two sugar monomers; sugars obtained from the hydrolysis of lignocellulosics contain a mixture of hexoses and pentoses making lignocellulose-based ethanol production a complex process [4, 19]. Thus, efficient economically viable ethanol production from lignocellulose requires conversion of all types of sugars produced from cellulose and hemicellulose [20, 21]. However, microorganisms have a natural preference towards the certain type of sugars they consume. Selection of suitable strains for fermentation is based on their response to inhibitors in fermenting media and different pretreatment methods. For instance, *Saccharomyces cerevisiae* JRC6 and *Candida tropicalis* JRC1 are recommended for alkali and acid pretreated substrates, respectively [22]. Various bacteria, fungi and yeast are able to utilize and ferment hexoses and pentoses to ethanol. Yeasts, especially *Saccharomyces cerevisiae*, also known as baker's yeast is the best-known microbe used for fermentation of hexoses such as glucose into ethanol due to its high ethanol productivity (18% (w/v)) and tolerance (150 g ethanol l^{-1}) [23]. These strains are also capable of fermenting galactose and mannose but only recombinant strains are known to ferment xylose and arabinose. Various types of hybrid, recombinant and wild-type yeast strains have been used in fermentation for ethanol production.

The ambient temperature for commercial strains of *Saccharomyces cerevisiae* is between 30 and 32 °C and their ethanol theoretical yields usually range between 90 and 93% [24]. Mukhtar et al. [25] reported that *Saccharomyces cerevisiae* also produces organic acids and acetaldehyde as by-products which can reduce the alcohol content by interacting with them during alcoholic fermentation.

The enzymatic hydrolysate (180 g/l) obtained from steam-exploded oak chips when fermented by *S. cerevisiae* produced an ethanol concentration, productivity and yield of 77 g/l, 16.9 g/l/h, 0.43 g/g [26]. Fermentation of enzymatic hydrolysate of alfaalfa with *S. cerevisiae* led to the consumption of 98% sugars with 85% fermentation efficiency and 1.3 g/l/h of ethanol productivity [27]. Fermentation of acid and alkali-treated cashew apple bagasse with *S. cerevisiae* produced 20.0 and 8.2 g/l ethanol with an ethanol productivity of 3.33 and 2.7 g/l/h, respectively [28]. In another study, enzymatic hydrolysate containing 110 g/l sugar when fermented with *S. cerevisiae* produced 45.7 g/l ethanol with an ethanol yield of 94% [29]. An ethanol yield of 0.48 g/g was achieved on fermentation of enzymatic hydrolysates of *P. juliflora* and *L. camera* containing 36.5 and 37.5 g/l sugars, respectively [30, 31]. The batch fermentation of sugarcane molasses was carried out using two commercial yeasts (New Aule Alcohol Yeast and New Aule baker's instant dry yeast) for 72 h with (0.3 vvm) and without aeration. New Aule Alcohol Yeast produced 74.8 g/L of ethanol with ethanol productivity and yield of 2.078 g/L/h and 0.378 g/g, respectively, and aeration did not exhibit any effect on ethanol production. New Aule baker's instant dry yeast produced 102.854 g/L of ethanol without aeration while aeration increased the ethanol production to 120.917 g/L with productivity and yield of 3.359 g/L/h and 0.669 g/g, respectively. This also indicated the differences in the oxygen sensitivity of both commercial yeasts [32]. Other yeast strains capable of fermenting different types of sugars to ethanol producers are *Pichia stipites* (NRRL-Y-7124), *S. cerevisiae* (RL-11) and *Kluyveromyces fragilis* (Kf1) [33]. In a study on isolation of potential fermenting strains, *Candida*, *Pichia* and *Dekkera* isolated from sugarcane molasses produced low concentrations of ethanol and acetic acid which act as an inhibitor of the fermentative yeast [34]. Also, some natural wild yeast species show the potential of replacing *S. cerevisiae* but their poor survival in the fermenter and low ethanol yield requires further research and improvement.

Bacterial strains are capable of using all sugars derived from cellulosic biomass for ethanol fermentation; however, catabolite repression is a limitation. *Zymomonas mobilis*, *Streptococcus fragilis* and *Kluyveromyces fragilis* are some of the widely used bacterial strains for commercial ethanol production [35]. *Z. mobilis* ferments glucose into ethanol and CO_2 through the ED pathway which generates more ethanol compared with the Embden–Meyerhof pathway in *S. cerevisiae*. *Z. mobilis* can also tolerate high concentrations of ethanol (120 g/L) compared to other fermenting organisms [36]. However, they have low tolerance to acetic acid and as little as 2.5 g/L of acetic acid can cause a decrease in ethanol productivity. This problem can be resolved by subjecting the hydrolysate to detoxification prior to the fermentation stage. The thermophilic bacteria like *Clostridium thermocellum* are known hydrolyze both hemicellulose and cellulose for ethanol production [37, 38]. However, production of mixed gaseous acidogenic fermentation products after cellulose hydrolysis and sensitivity to chemical inhibition is a major drawback [37, 39]. Some of the significant features of efficient fermenting organisms are broad substrate range; high ethanol yield, tolerance and productivity; minimal by-product formation; increased tolerance to inhibitors; tolerance to high temperature and osmotic pressure; recyclability and simultaneous hexose and pentose utilization [17, 19, 40].

7.3 Pentose Fermentation

Xylose is the most prominent pentose present in the hemicellulose of hardwood and crop residues followed by arabinose, and thus, their efficient utilization during fermentation is significant for cost reduction [41]. Also, pentose sugars produced on hydrolysis of lignocellulosic biomass create hindrance in yeast alcohol fermentation due to lack of xylose utilization enzymes like xylose reductase and xylitol dehydrogenase [21]. The yeast species capable of pentose fermentation are *Pichia stipitis (Scheffersomyces stipitis), Candida shehatae* and *Pachysolen tannophilus* [21, 42–44]. Among these, *S. stipitis* is widely known for its broad substrate specificity, broad range of sugars, high ethanol yield, the absence of xylitol production and no vitamin requirement for pentose utilization [44, 45]. Additionally, studies on utilization of strains of *S. stipitis* instead of traditional yeast have been conducted by a number of authors due to their fermenting capacity under anaerobic conditions with specific pentose sugar uptake at faster rates [46, 47]. Reports of bioethanol production on fermentation with *P. stipitis* from detoxified xylose-rich hydrolysates of *Prosopis juliflora* and *Lantana camara* are available [30, 31]. Similarly, fermentation of detoxified water hyacinth hemicellulose hydrolysate with *P. stipitis* NCIM-3497 resulted in ethanol yield of 0.425 g/g biomass [47]. *S. stipitis* has been reported to produce almost 50 g/L of ethanol from xylose with a theoretical fermentation capacity of 80% [44].

Other pentose fermenting microorganisms include fungal species like *Rhizopus oryzae, Fusarium oxysporum* and *Neurospora crassa* and bacterial species such as *Clostridium* sp., *Klebsiella* sp., *Lactobacillus* sp., *Bacillus macerans* DMS 1574, *Bacteroides polypragmatus* NRCC 2288 and *Erwinia chrysanthemi* B374 *Aeromonas hydrophila* [17, 21, 48–50]. However, employment of pentose fermenting organisms is limited due to the slow rate of fermentation, low ethanol yield, xylitol production, sensitivity to inhibitors including ethanol, strict oxygen requirement and by-product formation [4, 45]. *Scheffersomyces stipitis*, a pentose fermenting yeast is sensitive to organic acids and ethanol concentration above 30 g/L making it undesirable for industrial-scale use [51]. Co-culture fermentation provides an opportunity for simultaneous conversion of hexose and pentose along with increased ethanol yield and productivity [44]. Another possible solution is to use genetically modified strains of *S. cerevisiae* capable of utilizing pentoses. Also, pentoses could also be exploited for the production of other compounds at later stages in the process using filamentous fungi like *Rhizopus* sp. [52]. Use of food-related Zygomycetes for pentose fermentation has also been explored extensively, but the ethanol yield is most often limited (ca. 0.2 g/g) [49, 53, 54]. Metabolic engineering is another technique being employed to improve ethanol yield and productivity, broaden the substrate range, decreased by-product formation and introduction of new pathways for new product formation [4, 17]. Some of the organisms engineered to display attractive features for fermentation of lignocellulosics are *Z. mobilis*, *E. coli* and *S. cerevisiae* [4, 17, 19].

7.4 Bioethanol Production from Water Hyacinth

Water hyacinth (*Eichhornia crassipes*), a native of South America, is a noxious aquatic weed which is also widely found in tropical countries like India. It is responsible for the depletion of the various nutrients and oxygen content of the water, thereby affecting navigation, fishing and flora and fauna of the aquatic system. Under favourable conditions, water hyacinth can achieve a growth rate of up to 17.5 metric tonnes per hectare per day and population can double every 5–15 days [55]. Water hyacinth is low in lignin content (10%) and contains high amounts of hemicellulose (30–50%) and cellulose (20–40%) making it an attractive raw material [56]. Other favourable traits that make it potential biomass for bioethanol production are rapid proliferation, large availability, easily degradable and do not compete with arable crop plants for space, light and nutrients. Thus, ethanol production from water hyacinth biomass can significantly contribute to the weed management along with generation of employment and an income source in rural areas. Major fermentable sugars that present in water hyacinth hydrolysate are glucose and xylose [57]. Glucose fermentation to ethanol can be carried out by *S. cerevisiae* but xylose fermentation is challenging because of the shortage of efficient xylose fermenting microorganisms. This questions the economic viability of water hyacinth since the pentose sugars constitute up to 40% of water hyacinth biomass [44]. Xylose concentration of as high as 59% in water hyacinth hydrolysate has been reported [57]. Thus, for high ethanol productivity, any fermenting organism must be capable of fermenting all sugars and withstand potential inhibitors present in the water hyacinth hydrolysate [58].

Separated hydrolysis and fermentation (SHF) of enzymatic hydrolysate produced using a mixture of cellulase, xylanase and pectinase on dilute acid pretreated water hyacinth leaves was carried out using three different yeast strains, *Saccharomyces cerevisiae* TISTR 5048, *Saccharomyces cerevisiae* KM 1195 and *Saccharomyces cerevisiae* KM 7253. *Saccharomyces cerevisiae* KM 7253 resulted in the lowest yield of ethanol while *Saccharomyces cerevisiae* KM 1195 achieved the highest yields of ethanol. Co-culture of *S. cerevisiae* TISTR 5048 and *Candida tropicalis* TISTR 5045 further improved ethanol production with ethanol concentration, yield and productivity of 3.39 (g/l), 0.25 g/g and 0.221 g/l-h, respectively [59]. Another study on fermentation of acid pretreated water hyacinth by yeast (*Saccharomyces cerevisiae*) resulted in the production of 55.20% ethanol [60]. Bioethanol production from alkaline-oxidative pretreated water hyacinth was carried out by *Saccharomyces cerevisiae* KCTC 7928. Enzymatic saccharification was performed using Celluclast and Viscozyme L. Batch fermentation resulted in maximum ethanol yield of 0.35 g/g biomass [61]. In another study, a crude trial on fermentation of the enzymatic hydrolysate (cellulase and β-glucosidase) of alkali-treated water hyacinth biomass using the common baker's yeast *Saccharomyces cerevisiae* yielded an ethanol concentration of 4.4 g/L [62]. Separate hydrolysis and fermentation of water hyacinth biomass pretreated with *E. taxodii* (10 days) and 0.25% H_2SO_4 with *Saccharomyces cerevisiae* resulted in 0.192 g/g of dry matter of ethanol yield and was 1.34-fold higher than that from acid-treated water hyacinth. This indicated the inclusion of biological pretreatment as a promising method to improve enzymatic hydrolysis and ethanol production from water hyacinth [63].

Water hyacinth biomass after pretreatment with 4% H_2SO_4 was subjected to saccharification process by commercial cellulase enzyme (30 FPU/g biomass) and fermentation by *S. cerevisiae* leading to an ethanol concentration of 0.292% (w/v) with an overall process efficiency of 59.3% [64]. Ethanol production was performed using both mono- and co-culture of hexose and pentose fermenting yeasts. Ethanol yield using co-culture was 56% and 36% higher than monoculture of *S. cerevisiae* and *S. stipites,* respectively, due to the conversion of both hexose and pentose sugars to ethanol [55]. Simultaneous saccharification and fermentation of acid pretreated water hyacinth biomass were performed using 50 ml/L cellulase enzyme obtained from *A. niger* and *Zymomonas mobilis* CP4, a recombinant strain capable of hydrolyzing both hexoses and pentoses. An ethanol yield of 68.3 g/L was produced in 3.25 days [65]. Ganguly et al. [58] reported that fermentation of biologically pretreated water hyacinth biomass with *Pichia stipitis*, *Candida shehatae* and *Saccharomyces cerevisiae* resulted in ethanol yield of 3.496 g/L, 3.449 g/L and 3.144 g/L, respectively. Yeast fermentation of water hyacinth substrate after pretreatment with sulphuric acid and hydrolysis with cellulase enzyme (400 IU) was also studied under anaerobic conditions [66]. Enzymatic hydrolysate of dilute acid pretreated water hyacinth on fermentation with commercial baker's yeast (*Saccharomyces cerevisiae)* produced an ethanol yield of 0.484 g/g of reducing sugars which is 95% of theoretical yield (0.51 g/g glucose) [67].

Enzymatic hydrolysis of alkali pretreated water hyacinth was carried out by enzyme produced from *Aspergillus niger*. Enzymatic hydrolysate was subjected to concentration and fermentation by *Saccharomyces cerevisiae*, *Scheffersomyces stipitis* and co-culture of both, and ethanol concentrations of 4.3, 6.2 and 9.8 g/L were obtained [68]. The conversion of water hyacinth (*Eichhornia crassipes*) with 76.3% holocellulose and 12.3% lignin content into bioethanol using traditional strain of *S. cerevisiae* was studied. Continuous detoxification of hydrolysate using lime ("overliming") for degradation of furan compounds was also employed. An ethanol yield of 12.7 g/L comparable to those reported in other agricultural biomasses was produced [69]. Bioethanol production from alkali pretreated water hyacinth biomass was examined using both SHF and SiSF. For SHF, biomass was saccharified by crude enzymes of *T. reesei* and *F. oxysporum* and then fermented by *P. stipitis*. In SiSF, the process of saccharification and fermentation was carried out using both mono and co-cultures of selected fungal strains. The highest ethanol yield of 0.411 g g^{-1} was obtained from co-culture of *T. reesei* and *P. stipites* as a result of the ability of *P. stipitis* to ferment pentose sugars to ethanol [70].

The potential of four ethanol producing yeast strains, namely *Pachysolen tannophilus, Candida intermedia, Pichia stipitis* and *Saccharomyces cerevisiae,* was explored for fermentation of acid pretreated water hyacinth biomass. Enzymatic hydrolysis was done using the enzyme complex (cellulase 18.33 IU/mL and xylanase 31.43 IU/mL) produced from *Trichoderma reesei* NRRL-3652. The highest ethanol production was obtained with *P. tannophilus* reaching a maximum ethanol concentration of 0.043 g/g, followed by 0.021, 0.037 and 0.015 g/g for *C. intermedia, P. stipites* and *S. cerevisiae* [71]. In a study, fermentation experiments of enzymatic hydrolysate of alkali pretreated water hyacinth obtained on hydrolysis with 49.56 U/g cellulase and 280.33 U/g xylanase were performed in shake flasks using *Candida shehatae*, *Saccharomyces cerevisiae* and *Pichia stipitis*. A highest ethanol yield of 0.1044 g/g dry substrate was obtained with *Pichia stipitis*, followed by 0.0824 g/g and 0.0676 g/g from *Candida shehatae* and *S. cerevisiae,* respectively. Highest ethanol yield from *P. Stipitis* can be explained by the fact that it can utilize both pentose and hexose sugar to produce ethanol. However, water hyacinth is rich in both cellulose and hemicellulose; *P. stipitis* produced a higher amount of ethanol [72]. In another work by Das et al. [73], fermentation process of acid pretreated water hyacinth biomass on enzymatic hydrolysis with cellulase enzyme (30 FPU) was optimized using central composite design and an ethanol yield of 13.6 mg/ml was obtained on co-fermentation by *Saccharomyces cerevisiae* (MTCC 173) and *Zymomonas mobilis* (MTCC 2428). Water hyacinth was used as a substrate for bioethanol production using *Saccharomyces cerevisiae* by SSF process after acid pretreatment and enzymatic hydrolysis by cellulase extracted from *Trichoderma viride* and 1.289 g/L of ethanol was produced [74]. The yield of ethanol is greatly influenced by the presence of inhibitory products like furfural or hydroxyl methyl furfurals which may be present in the hydrolysate. These inhibitory substances inhibit the central enzymes like hexokinase, phosphofructokinase and triose-phosphate dehydrogenase required for ethanol production. These studies represent the suitability of water hyacinth, a conventional invasive weed as a potential feedstock for bioethanol energy production.

7.5 Recent Research and Recombinant Fermentative Microbes

Ethanol yield and productivity are an important parameter to determine the economic viability of bioethanol production because of the cost of the raw material, pretreatment and cellulolytic enzymes which constitute a major part of the total production cost. Selection and improvement of industrial strains are a feasible approach for the generation of new strains to be used as a platform to produce second-generation ethanol and a critical step towards enabling economic biofuel production [75]. Wild fermenting microorganisms do not meet the requirements for industrial production of second-generation bioethanol, and thus, research on synthetic biology and metabolic engineering to improve them has been the focus. Significant features of focus for potential industrial strains include high tolerance to elevated temperatures, inhibitors, and ethanol, enzymes recycling and high sugar concentrations, efficient growth in lignocellulosic hydrolysate, effective xylose utilization and co-fermentation of pentoses and hexoses. Major limitation of wild microorganisms is their inability to produce high ethanol yields due to the lack of enzymes for both pentoses and hexoses. Genetic modification is thus crucial for enhancing the range of consumed sugars for cost-effective ethanol production. Synthetic biology and metabolic engineering have contributed to co-fermentation of hexoses and pentoses either by mutagenesis or introduction of a heterologous metabolic pathway for xylose utilization in conventional strain and improved ethanol productivity and yield. Besides, the use of recombinant DNA technology to upregulate stress tolerance genes has also been explored to overcome inhibitory situations [76]. *S. cerevisiae* has been successfully engineered to ferment glucose and pentose sugars simultaneously. *S. cerevisiae* cells were subjected to lithium acetate transformation method to obtain three recombinants, namely HDY-ZMYWBG1, HDY-ZMYWBG2 and HDY-ZMYWBG3. The resulting consortium demonstrated to utilize both pentose and hexoses, and the final ethanol production was threefold higher than a similar yeast consortium secreting only the three cellulases [77, 78]. Similarly, another study reported generation of engineered yeasts to co-ferment mixtures of xylose and cellobiose resulting in improved ethanol yield [79]. A major limitation associated with recombinant yeasts sometimes is that they do not ferment hexoses and pentoses simultaneously leading to a prolonged fermentation period. Pentose utilization begins after complete ethanol conversion of glucose due to the stronger affinity of glucose for transporters than xylose [80]. Also, the amount of pentose utilized by recombinant yeasts is relatively lower as compared to hexose during fermentation making the process of ethanol production uneconomical for commercial scale. Although metabolic engineering has contributed significantly in this area, more researches to develop genetic modifications techniques, to develop yeast strains capable of co-fermenting pentoses and hexoses and to produce reaction intermediates and efficient pentose transporter for the production of lignocellulosic ethanol is required [40, 81].

Another strategy for efficient and economical production of bioethanol from lignocellulosic feedstock is by introducing cellulase-encoding genes into specific species to reduce the use of exogenous cellulases during the process [82]. However, a study reported that the application of exogenous cellulases was still found to be a requisite even on the introduction of cellulase-encoding genes into *S. cerevisiae* [79]. Cellodextrin transport system from *N. crassa* was reconstituted in *Saccharomyces cerevisiae* to promote better growth on cellodextrin resulting in higher ethanol productivity in engineered yeast cells compared to wild-type cells [79, 83]. Use of thermotolerant microbes to achieve ethanol fermentation at high temperature is an emerging technology to maintain the temperature imbalance during SiSF and CBP processes [84]. *K. marxianus* has been a potential candidate to be exploited in this technology for its ability to co-ferment both hexose and pentose sugars and survive high-temperature range of 42–45 °C [85]. *K. marxianus* has also been genetically modified to exhibit cellulolytic activities from *T. reesei* and *Aspergillus aculeatus* allowing uninterrupted conversion of lignocellulose into ethanol during CBP process at 48 °C producing 0.47 g/g ethanol [85].

Genetic engineering to achieve simultaneous conversion of hexoses and pentoses for ethanol production from bacteria has mainly been focused on the gram-negative bacteria *Escherichia coli, Zymomonas mobilis* and *Klebsiella oxytoca* and gram-positive bacteria *Lactobacillus casei* and *Clostridium cellulolyticum* [40]. *Z. mobilis* is known for its high ethanol yield and resistance to high-temperature range. Report of improvement in ethanol productivity from fermenting bacteria such as *E. coli* and *Z. mobilis* using genetic modifications is available [17, 86, 87]. *Z. mobilis* has been an attractive candidate for genetic modification, and numerous genes have been introduced to extend its substrate range for effective utilization of sugars like xylose and arabinose [17, 87]. Recombinant microorganisms capable of metabolizing multiple sugars can further be modified to produce bioethanol through the glycolysis pathway [88]. *S. cerevisiae* has already established its application in first-generation large-scale industrial bioethanol production. However, its industrial application in second-generation bioethanol production, especially CBP, is gaining increasing research. Recombinant stains of both *Z. mobilis* and *S. cerevisiae* have shown higher alcohol tolerance and conversion efficiency [89, 90]. Thus, recent advances in genetically engineering of *S. cerevisiae* and *Z. mobilis* are promising for the replacement of the conventional first-generation bioethanol by second-generation bioethanol provided current processing limitations are overcome and the best combination of advanced production systems is used to generate higher alcohol yields.

Assessment of the economic feasibility of bioethanol plant requires analysis of factors like type of feedstock and its handling, pretreatment method, cellulolytic enzymes production/purchase and hydrolysis, concentration of hexoses and pentoses, fermentation/co-fermentation techniques, distillation and anaerobic digestion, carbon dioxide recovery and furfural production. A six-step cost evaluation analysis to design lignocellulosic ethanol pathway has also been proposed and involves the following steps: (1) identification of desired ethanol characteristics, (2) to target cost of lignocellulosic ethanol, (3) to target expense of the supply pathway, (4) to decide on selling price of ethanol, (5) determination of cost management strategies, (6) constant tryout for advancement [91]. Research institutions all over the globe have made significant contributions to the techno-economic evaluation of bioethanol production [92–94]. Techno-economic evaluation in terms of resource potential, pretreatment processes, conversion rate of different lignocellulosics and different fermentation strategies like SiSF and CBP have been investigated [91, 93]. These different studies on the economic evaluation of ethanol production from lignocellulosic materials suggest that the annual production costs of bioethanol production vary significantly in the range of USD 0.13–0.87/L of ethanol [90–94].

References

1. IEA (2010) Sustainable production of second-generation biofuels. Potential and perspectives in major economies and developing countries. (http://www.iea.org/publications/freepublications/publication/biofuels_exec_summary.pdf)
2. Sanchez OJ, Cardona CA (2008) Trends in biotechnological production of fuel ethanol from different feedstocks. Bioresour Technol 99:5270–5295
3. Sun Y, Cheng J (2002) Hydrolysis of lignocellulosic materials for ethanol production: a review. Bioresour Technol 83(1):1–11
4. Hahn-Hagerdal B, Galbe M, Gorwa-Grauslund MF, Liden G, Zacchi G (2006) Bio-ethanol-the fuel of tomorrow from the residues of today. Trends Biotechnol 24(12):549–556
5. Sukumaran RK, Singhania RR, Mathew GM, Pandey A (2009) Cellulase production using biomass feed stock and its application in lignocellulose saccharification for bio-ethanol production. Renew Energy 34(2):421–424
6. Cotanaa F, Cavalaglioa G, Gelosiaa M, Nicolinia A, Cocciaa V, Petrozzia A (2014) Production of bioethanol in a second generation prototype from pine wood chips. Energy Procedia 45:42–51
7. Gauss WF, Suzuki S, Takagi (1976) Manufacture of alcohol from cellulosic materials using plural ferments. Ed: Office USPT. USA, Bio Research Center Company Limited 3990944(610731)
8. Cong Y (2010) Bioethanol production from microwave-pretreated and hydrolyzed water hyacinth by fermentation. Master's thesis—Dissertation Zhejiang University
9. Ohgren K, Bura R, Lesnicki G, Saddler J, Zacchi G (2007) A comparison between simultaneous saccharification and fermentation and separate hydrolysis and fermentation using steam-pretreated corn stover. Proc Biochem 42:834–839
10. Philippidis GP, Smith TK (1995) Limiting factors in the simultaneous saccharification and fermentation process for conversion of cellulosic biomass to fuel ethanol. Appl Biochem Biotechnol 51–52:117–124
11. Taherzadeh MJ, Karimi K (2007) Enzyme based hydrolysis process for ethanol from lignocellulosic materials: a review. BioResources 2(4):707–738
12. Zhang J, Lynd LR (2010) Ethanol production from paper sludge by simultaneous saccharification and co-fermentation using recombinant xylose-fermenting microorganisms. Biotechnol Bioeng 107:235–244
13. Kim TH, Lee YY, Sunwoo C, Kim JS (2006) Pretreatment of corn stover by low liquid ammonia recycle percolation process. Appl Biochem Biotech 133(1):41–57
14. Hamelinck CN, van Hooijdonk G, Faaij APC (2005) Ethanol from lignocellulosic biomass: techno-economic performance in short-middle- and long term. Biomass Bioenergy 28:384–410
15. Klein-Marcuschamer D, Oleskowicz-Popiel P, Simmons BA, Blanch HW (2010) Technoeconomic analysis of biofuels: a wiki based platform for lignocellulosic biorefineries. Biomass Bioenerg 34(12):1914–1921
16. Szczodrak J, Fiedurek J (1996) Technology for conversion of lignocellulosic biomass to ethanol. Biomass Bioenerg 10:367–375
17. Dien BS, Cotta MA, Jeffries TW (2003) Bacteria engineered for fuel ethanol production: current status. Appl Microbiol Biotechnol 63:258–266
18. Sreenath HK, Jeffreis TW (2000) Production of ethanol from wood hydrolysate by yeasts. Bioresour Technol 72:253–260
19. Zaldivar J, Nielsen J, Olsson L (2001) Fuel ethanol production from lignocellulose: a challenge for metabolic engineering and process integration. Appl Microbiol Biotechnol 56:17–34
20. Jeffries TW (2006) Engineering yeasts for xylose metabolism. Curr Opin Biotechnol 17: 320–326
21. Hahn-Hagerdal B, Karhumaa K, Jeppsson M, Gorwa-Grauslund MF (2007) Metabolic engineering for pentose utilization in *Saccharomyces cerevisiae*. Adv Biochem Eng Biotechnol 108:147–177

22. Choudhary J, Singh S, Nain L (2017) Bioprospecting thermotolerant ethanologenic yeasts for simultaneous saccharification and fermentation from diverse environments. J Biosci Bio-eng 123(3):342–346
23. Claassen PAM, van Lier JB, Lopez Contreras AM, van Niel EWJ, Sijtsma L, Stams AJM, de Vries SS, Weusthuis RA (1999) Utilisation of biomass for the supply of energy carriers. Appl Microbiol Biotechnol 52:741–755
24. Thammasittirong SN, Thirasaktana T, Thammasittirong A, Srisoduk M (2013) Improvement of ethanol production by ethanol-tolerant *Saccharomyces cerevisiae* UVNR56. Springer Plus 2:583
25. Mukhtar K, Asgher M, Afghan S, Hussain K, Zia-ul-Hussain S (2010) Comparative study on two commercial strains of *Saccharomyces cerevisiae* for optimum ethanol production on industrial scale. J Biomed Bioechnol 1–5
26. Lee YY, Iyer P, Torget RW (1999) Dilute-acid hydrolysis of lignocellulosic biomass. Adv Biochem Eng Biotechnol 65:93–115
27. Belkacemi K, Turcotte G, De Halleux D, Savoie P (1997) Ethanol production from AFEX-treated forages and agricultural residues. Making a business from biomass in energy, environment, chemicals, fibres and materials. In: Proceedings of the 3rd biomass conference of the Americas, Monteral, pp 1117–1130
28. Rocha MVP, Rodrigues TH, de Macedo GR, Goncalves LR (2009) Enzymatic hydrolysis and fermentation of pretreated cashew apple bagasse with alkali and diluted sulfuric acid for bioethanol production. Appl Biochem Biotechnol 155:407–417
29. Chen Y, Sharma-Shivappa RR, Chen C (2007) Ensiling agricultural residues for bioethanol production. Appl Biochem Biotechnol 143:80–92
30. Gupta R, Sharma KK, Kuhad RC (2009) Separate hydrolysis and fermentation (SHF) of Prosopis juliflora, woody substrate for the production of cellulosic ethanol by *Saccharomyces cerevisiae* and Pichia stipitis-NCIM 3498. Bioresour Technol 100:1214–1220
31. Kuhad RC, Gupta R, Khasa YP, Singh A (2010) Bioethanol production from Lantana camera (red sage): pretreatment, saccharification and fermentation. Bioresour Technol 101:8348–8354
32. Jayusa Nurhayatia, Mayzuhroha A, Arindhania S, Caroenchaic C (2016) Studies on bioethanol production of commercial baker's and alcohol yeast under aerated culture using sugarcane molasses as the media. Agricult Sci Procedia 9:493–499
33. Mussatto SI, Dragone G, Guimaraes PMR, Silva JPA, Carneiro LM, Roberto IC, Vicente A, Domingues L, Teixeira JA (2010) Technological trends, global market and challenges of bio-ethanol production-R1. Biotechnol Adv 28:817–830
34. Basilio ACM, Araujo PRL, Morais JOF, da Silva Filho EA, de Morais Jr MA, Simoes DA (2008) Detection and identification of wild yeast contaminants of the industrial fuel ethanol fermentation process. Cur Microbiol 56(4):322–326
35. Pesta G, Meyer-Pittroff R, Russ W (2006) Utilization of whey. In: Oreopoulou V, Russ W (eds) Utilization of byproducts and treatment of waste in the food industry, vol 1. Springer, New York, pp 1–11
36. Zhao XQ, Zi LH, Bai FW, Lin HL, Hao XM, Yue GJ, Ho NWY (2011) Bioethanol from lignocellulosic biomass. Adv Biochem Eng/Biotechnol 128:25–51
37. Lynd LR, Weimer PJ, van zyl WH WH, Pretorius IS (2002) Microbial cellulose utilization: fundamentals and biotechnology and microbiology. Mol Biol Rev 66:506–577
38. Wu KJ, Saratale GD, Lo YC, Chen SD, Chen WM, Tseng ZJ, Chang JS (2008) Fermentative production of 2, 3 butanediol, ethanol and hydrogen with *Klebsiella* sp. isolated from sewage sludge. Bioresour Technol 99:7966–7970
39. Demain AL, Newcomb M, Wu JH (2005) Cellulase, *clostridia* and ethanol. Microbiol Mol Biol Rev 69:124–154
40. Senthilkumar V, Gunasekaran P (2005) Bioethanol production from cellulosic substrates: engineered bacteria and process integration challenges. J Sci Ind Res 64:845–853
41. Prasad S, Singh A, Joshi HC (2007) Ethanol as an alternative fuel from agricultural, industrial and urban residues. Resour Conserv Recycl 50:1–39

42. Mosier NS, Wyman C, Dale B, Elander R, Lee YY, Holtzapple M, Ladisch MR (2005) Features of promising technologies for pretreatment of lignocellulosic biomass. Bioresour Technol 96:673–686
43. Kuhad RC, Gupta R, Khasa YP, Singh A (2011) Bioconversion of pentose sugars to ethanol: current and future prospects. Renew Sustain Energy Rev 15:4950–4962
44. Chen Y (2011) Development and application of co-culture for ethanol production by co-fermentation of glucose and xylose: a systematic review. J Ind Microbiol Biotechnol 38:581–597
45. Du Preez JC, Bosch M, Prior BA (1986) The fermentation of hexose and pentose sugars by *Candida shehatae* and *Pichia stipitis*. Appl Microbiol Biotechnol 23:228–233
46. Bhattacharya A, Ganguly A, Das S, Chatterjee PK, Dey A (2013) Fungal isolates from local environment: isolation, screening and application for the production of ethanol from water hyacinth. Int J Emerg Tech Adv Engi 3(3):58–65
47. Kumari N, Bhattacharya A, Dey A, Ganguly A, Chatterjee PK (2014) Bioethanol production from water hyacinth biomass using isolated fungal strain from local environment. Biolife 2(2):516–522
48. Chaillou S, Pouwels PH, Postma PW (1999) Transport of D-xylose in *Lactobacillus pentosus*, *Lactobacillus casei* and *Lactobacillus plantarum*: Evidence for a mechanism of facilitated diffusion via the phosphoenolpyruvate: mannose phosphotransferase system. J Bacteriol 181:4768–4773
49. Millati R, Edebo L, Taherzadeh MJ (2005) Performance of *Rhizopus*, *Rhizomucor* and *Mucor* in ethanol production from glucose, xylose and wood hydrolysates. Enz Microb Technol 36:294–300
50. Ruiz E, Romero I, Moya M, Sanchez S, Bravo V, Castro E (2007) Sugar fermentation by *Fusarium oxysporum* to produce ethanol. World J Microbiol Biotechnol 23:259–267
51. Agbobo FK, Coward-Kelly G, Torry-Smith M, Wenger K, Jeffries TW (2007) The effect of initial cell concentration on xylose fermentation by *Pichia stipitis*. Appl Biochem Biotechnol 137–140(1–12):653–662
52. Taherzadeh MJ, Fox M, Hjorth H, Edebo L (2003) Production of mycelium biomass and ethanol from paper pulp sulfite liquor by *Rhizopus oryzae*. Bioresour Technol 88(3):167–177
53. Ferreira JA, Lennartsson PR, Niklasson C, Lundin M, Edebo L, Taherzadeh MJ (2012) Spent sulphite liquor for cultivation of an edible *Rhizopus* sp. BioResources 7(1):173–188
54. Wikandari R, Millati R, Lennartsson PR, Harmayani E, Taherzadeh MJ (2012) Isolation and characterization of zygomycetes fungi from tempe for ethanol production and biomass applications. Appl Biochem Biotechnol 167(6):1501–1512
55. Singh A, Bishnoi NR (2012) Optimization of ethanol production from microwave alkali pretreated rice straw using statistical experimental designs by *Saccharomyces cerevisiae*. Ind Crop Prod 37:334–341
56. Bhattacharya A, Kumar P (2010) Water hyacinth as a potential biofuel crop. Electron J Environ Agricult Food Chem 9:112–122
57. Nigam JN (2002) Bioconversion of water-hyacinth (*Eichhornia crassipes*) hemicellulose acid hydrolysate to motor fuel ethanol by xylose-fermenting yeast. J Biotechnol 97:107–116
58. Ganguly A, Das S, Bhattacharya A, Dey A, Chatterjee PK (2013) Enzymatic hydrolysis of water hyacinth biomass for the production of ethanol: optimization of driving parameters. Indian J Exp Bio 51(7):556–566
59. Sornvoraweat B, Kongkiattikajorn J (2010) Separated hydrolysis and fermentation of water hyacinth leaves for ethanol production. Khon Kaen University Research Journal 15:794–802
60. Mahmood T, Malik SA, Hussain ST (2010) Biosorption and recovery of heavy metals from aqueous solutions by *Eichhornia crassipes* (water hyacinth) ash. BioResources 5:1244–1256
61. Ahn DJ, Kim SK, Yun HS (2012) Optimization of pretreatment and saccharification for the production of bioethanol from water hyacinth by *Saccharomyces cerevisiae*. Bioproc Biosys Eng 35(1–2):35–41
62. Aswathy US, Rajeev K, Sukumaran G, Lalitha Devi KP, Rajasree R, Singhania R, Pandey A (2010) Bio-ethanol from water hyacinth biomass: an evaluation of enzymatic saccharification strategy. Bioresour Technol 101(3):925–930

63. Ma F, Yang N, Xu C, Yu H, Wu J, Zhang X (2010) Combination of biological pretreatment with mild acid pretreatment for enzymatic hydrolysis and ethanol production from water hyacinth. Bioresour Technol 101(24):9600–9604
64. Satyanagalakshmi K, Sindhu R, Binod P, Janu KU, Sukumaran RK, Pandey A (2011) Bioethanol production from acid pretreated water hyacinth by separate hydrolysis and fermentation. J Sci Ind Res 70:156–161
65. Kasthuri T, Gowdhaman D, Ponnusami V (2012) Production of ethanol from water hyacinth (*Eichhornia crassipes*) by *Zymomonas mobilis* CP4: optimization studies. Asian J Sci Res 5:285–289
66. Takagi T, Uchida M, Matsushima R, Ishida M, Urano N (2012) Efficient bioethanol production from water hyacinth *Eichhornia crassipes* by both preparation of the saccharified solution and selection of fermenting yeasts. Fisheries Sci 78(4):905–910
67. Idreesa M, Adnana A, Sheikhb S, Qureshi FA (2013) Optimization of dilute acid pretreatment of water hyacinth biomass for enzymatic hydrolysis and ethanol production. EXCLI J 12:30–40
68. Singh A, Bishnoi NR (2013) Comparative study of various pretreatment techniques for ethanol production from water hyacinth. Ind Crops Prod 44:283–289
69. Caparanga AR and Bathan RS (2013) Bioconversion of water hyacinth (*Eichhornia crassipes*) at its optimum age into bioethanol via separate hydrolysis and fermentation with continuous overliming. Annual International congress on energy, San Francisco, CA Nov 3–8, No. 342407
70. Pothiraj C, Arumugam R, Gobinath M (2014) Sustaining ethanol production from lime pretreated water hyacinth biomass using mono and co-cultures of isolated fungal strains with *Pichia stipites*. Bioresour Bioproc 1:27
71. Manivannan A, Narendhirakannan RT (2015) Bioethanol production from aquatic weed water hyacinth (*Eichhornia crassipes*) by yeast fermentation. Waste Biomass Valor 6:209–216
72. Das S, Bhattacharya A, Haldar S, Ganguly A, Glu S, Ting YP, Chatterjee PK (2015) Optimization of enzymatic saccharification of water hyacinth biomass for bioethanol: comparison between artificial neural network and RSM. Sustain mater Technol 3:17–28
73. Das A, Ghosh P, Paul T, Ghosh U, Pati BR, Mondal KC (2016) Production of bioethanol as useful biofuel through the bioconversion of water hyacinth (*Eichhornia crassipes*). 3Biotech 6:70
74. Zhang Q, Weng C, Huang H, Achal V, Wang D (2016) Optimization of bioethanol production using whole plant of water hyacinth as substrate in simultaneous saccharification and fermentation process. Front Microbiol 6:1411
75. Lopes ML, de Lima Paulillo SC, Godoy A, Cherubin RA, Lorenzi MS, Giometti FHC, Bernardino CD, de Amorim Neto HB, de Amorim HV (2016) Ethanol production in Brazil: a bridge between science and industry. Brazil J Microbiol 47S:64–76
76. Dogan A, Demirci S, Aytekin AO, Sahin F (2014) Improvements of tolerance to stress conditions by genetic engineering in *Saccharomyces cerevisiae* during ethanol production. Appl Biochem Biotechnol 174(1):28–42
77. Ge JP, Zhang LY, Ping WX, Zhang MY, Shen Y, Song G (2014) Genetically engineered *Saccharomyces cerevisiae* strain that can utilize both xylose and glucose for fermentation. Appl Mechan Mat 448–453:1637–1643
78. Goyal G, Tsai SL, Madan B, DaSilva NA, Chen W (2011) Simultaneous cell growth and ethanol production from cellulose by an engineered yeast consortium displaying a functional mini-cellulosome. Microb Cell Factor 10:89
79. Ha SJ, Galazka JM, Kim SR et al (2011) Engineered *Saccharomyces cerevisiae* capable of simultaneous cellobiose and xylose fermentation. Proc Natl Acad Sci USA 108(2):504–509
80. Oreb M, Dietz H, Farwick A, Boles E (2012) Novel strategies to im-prove co-fermentation of pentoses with d-glucose by re-combinant yeast strains in lignocellulosic hydrolysates. Bioengineered 3(6):347–351
81. Robak K, Balcerek M (2018) Review of second generation bioethanol production from residual biomass. Food Technol Biotechnol 56(2):174–187
82. Kricka W, Fitzpatrick J, Bond U (2014) Metabolic engineering of yeasts by heterologous enzyme production for degradation of cellulose and hemicellulose from biomass: a perspective. Front Microbiol 5:174

83. Galazka JM, Tian C, Beeson WT, Martinez B, Glass NL, Cate JHD (2010) Cellodextrin transport in yeast for improved biofuel production. Science 330(6000):84–86
84. Fonseca GG, Heinzle E, Wittmann C, Gombert AK (2008) The yeast *Kluyveromyces marxianus* and its biotechnological potential. Appl Microbiol Biotechnol 79(3):339–354
85. Yanase S, Hasunuma T, Yamada R et al (2010) Direct ethanol production from cellulosic materials at high temperature using the thermotolerant yeast *Kluyveromyces marxianus* displaying cellulolytic enzymes. Appl Microbiol Biotechnol 88(1):381–388
86. Dien BS, Iten LB, Bothast RJ (1999) Conversion of corn fiber to ethanol by recombinant *E. coli* strain FBR3. J Ind Microbiol Bio-technol 22(6):575–581
87. Deanda K, Zhang M, Eddy C, Picataggio S (1996) Development of an arabinose-fermenting *Zymomonas mobilis* strain by metabolic pathway engineering. Appl Environ Microbiol 62(12):4465–4470
88. Sarris D, Papanikolaou S (2016) Biotechnological production of ethanol: biochemistry, processes and technologies. Eng Life Sci 16(4):307–329
89. Weber C, Farwick A, Benisch F et al (2010) Trends and challenges in the microbial production of lignocellulosic bioalcohol fuels. Appl Microbiol Biotechnol 87(4):1303–1315
90. Kang Q, Appels L, Tan T, Dewil R (2014) Bioethanol from lignocellulosic biomass: current findings determine research priorities. Sci World J. https://doi.org/10.1155/2014/298153
91. Gnansounou E, Dauriat A (2010) Techno-economic analysis of lignocellulosic ethanol: a review. Bioresour Technol 101:4980–4991
92. Aden A, Foust T (2009) Techno-economic analysis of the dilute sulfuric acid and enzymatic hydrolysis process for the conversion of corn stover to ethanol. Cellulose 16:535–545
93. Sassner P, Galbe M, Zacchi G (2008) Techno-economic evaluation of bioethanol production from three different lignocellulosic materials. Biomass Bioener 32:422–430
94. Humbird D, Aden A (2009) Biochemical production of ethanol from corn stover: 2008 State of Technology Model National Renewable Energy Laboratory (NREL) 2009. NREL Report TP-510–46214

MIX
Papier aus verantwortungsvollen Quellen
Paper from responsible sources
FSC® C105338

If you have any concerns about our products,
you can contact us on
ProductSafety@springernature.com

In case Publisher is established outside the EU,
the EU authorized representative is:
Springer Nature Customer Service Center GmbH
Europaplatz 3, 69115 Heidelberg, Germany

Printed by Libri Plureos GmbH
in Hamburg, Germany